OCR

Chemistry A

Modules 3 and 4

Periodic table and energy

Core organic chemistry

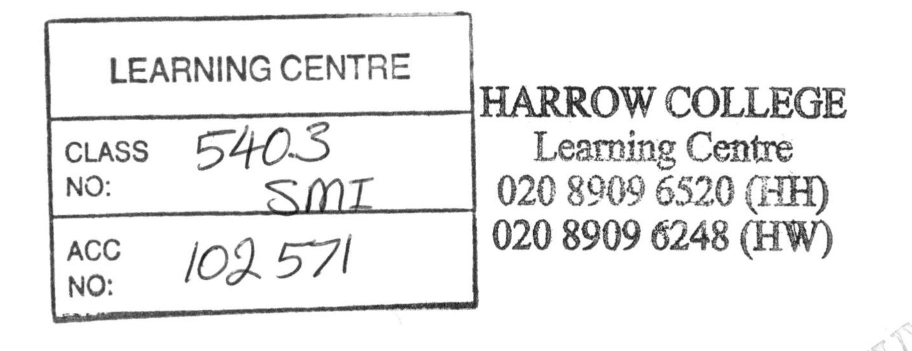

Mike Smith

PHILIP ALLAN FOR HODDER EDUCATION
AN HACHETTE UK COMPANY

Philip Allan, an imprint of Hodder Education, an Hachette UK company, Blenheim Court, George Street, Banbury, Oxfordshire OX16 5BH

Orders

Bookpoint Ltd, 130 Milton Park, Abingdon, Oxfordshire OX14 4SB

tel: 01235 827827

fax: 01235 400401

e-mail: education@bookpoint.co.uk

Lines are open 9.00 a.m.–5.00 p.m., Monday to Saturday, with a 24-hour message answering service. You can also order through the Hodder Education website: www.hoddereducation.co.uk

ISBN 978-1-4718-4399-0

First printed 2015

Impression number 5 4 3 2 1

Year 2018 2017 2016 2015

This guide has been written specifically to support students preparing for the OCR AS and A-level Chemistry examinations. The content has been neither approved nor endorsed by OCR and remains the sole responsibility of the author.

Cover photo: Galyna Andrushko/Fotolia

Typeset by Integra Software Services Pvt Ltd, Pondicherry, India

Printed in Italy

Hachette UK's policy is to use papers that are natural, renewable and recyclable products and made from wood grown in sustainable forests. The logging and manufacturing processes are expected to conform to the environmental regulations of the country of origin.

Contents

Content Guidance

Questions & Answers

Getting the most from this book

Exam tips

Advice on key points in the text to help you learn and recall content, avoid pitfalls, and polish your exam technique in order to boost your grade.

Knowledge check

Rapid-fire questions throughout the Content Guidance section to check your understanding.

Knowledge check answers

1 Turn to the back of the book for the Knowledge check answers.

Summaries

- Each core topic is rounded off by a bullet-list summary for quick-check reference of what you need to know.

Exam-style questions

Commentary on the questions

Tips on what you need to do to gain full marks, indicated by the icon ⓔ

Sample student answers

Practise the questions, then look at the student answers that follow.

Commentary on sample student answers

Find out how many marks each answer would be awarded in the exam and then read the comments (preceded by the icon ⓔ) showing exactly how and where marks are gained or lost.

Questions & Answers

(a) Explain what is meant by the term activation energy. (1 mark)

(b) On the sketch above, mark E_c, the activation energy in the presence of a catalyst. (1 mark)

(c) Explain, in terms of the distribution curve, how a catalyst speeds up the rate of a reaction. (2 marks)

(d) (i) Raising the temperature can also increase the rate of this reaction. Draw a second curve to represent the energy distribution at a higher temperature. Label your curve T_2. (2 marks)

(ii) Explain how an increase in temperature can speed up the rate of a reaction. (2 marks)

Total: 8 marks

ⓔ The command word 'explain' in (a), (c) and (d)(ii) infers that reasons are needed to support your statements. The command word 'draw' in (d)(i) is self-explanatory and a diagram is required.

Student A

(a) The activation energy is the minimum energy needed to start a reaction.

Student B

(a) The energy needed for a collision to be successful.

ⓔ Student A gains the mark, but Student B doesn't. The key word missing from Student B's response is *minimum*.

Student A

(b)

Student B

(b)

ⓔ Both score the mark. The activation energy for the catalyst must be lower than the original activation energy.

72 OCR Chemistry A

About this book

This guide is the second of a series covering the OCR AS Chemistry A (H032) and the OCR A Chemistry A (H432) specifications. It offers advice for the effective study of Modules 3 and 4.

The **Content Guidance** gives a point-by-point description of all the facts you need to know and concepts that you need to understand for Modules 3 and 4. It aims to provide you with a basis for your revision. However, you must also be willing to use other sources in your preparation for the examination.

Module 3, Periodic table and energy, includes basic information that you will need for the AS/A-level course. The energy section is particularly important as this will be built upon and extended in the second year of the course.

Module 4, Core organic chemistry, includes basic information that you will need for the AS/A-level course. A good understanding of this module, particularly the understanding of mechanisms and characteristic reactions of different functional groups, provides you with a solid foundation on which to build.

Practical skills are embedded throughout all the modules and will be assessed in the written examinations.

Modules 3 and **4** feature in both AS examinations.

Module 3 will be examined in papers 1 and 3 of the A-level course and Module 4 will be tested in papers 2 and 3.

The **Questions & Answers** section shows you the sort of questions you can expect in the component tests. It would be impossible to give examples of every kind of question in one book, but the questions used should give you a flavour of what to expect. Unless otherwise stated the questions are relevant to both AS and A-level students. Each question has been attempted by two students, Student A and Student B. Their answers, along with the comments, should help you see what you need to do to score a good mark — and how you can easily *not* score marks, even though you probably understand the chemistry.

What can I assume about the guide?

You can assume that:

- the topics covered in the Content Guidance relate directly to those in the specification
- the basic facts you need to know are stated clearly
- the major concepts you need to understand are explained
- the questions at the end of the guide are similar in style to those that will appear in the component tests
- the answers supplied are genuine, combining responses commonly written by students
- the standard of the marking is broadly equivalent to the standard that will be applied to your answers

What can I *not* assume about the guide?

You must *not* assume that:

- every last detail has been covered
- the way in which the concepts are explained is the *only* way in which they can be presented in an examination (often concepts are presented in an unfamiliar situation)
- the range of question types presented is exhaustive (examiners are always thinking of new ways to test a topic)

Study skills and revision techniques

All students need to develop good study skills. This section provides advice and guidance on how to study AS and first year A-level chemistry.

Organising your notes

Chemistry students often accumulate a large quantity of notes, so it is useful to keep these in a well-ordered and logical manner. It is necessary to review your notes regularly, maybe rewriting the notes taken during lessons so that they are clear and concise, with key points highlighted. You should check your notes using textbooks, and fill in any gaps. Make sure that you go back and ask your teacher if you are unsure about anything, especially if you find conflicting information in your class notes and textbook.

It is a good idea to file your notes in specification order using a consistent series of headings. The Content Guidance can help you with this.

Organising your time

Preparation for examinations is personal. Different people prepare, equally successfully, in different ways. The key is being honest about what *works for you*.

Whatever your style, you must have a plan. Sitting down the night before the examination with your notes and a textbook does not constitute a revision plan — it is just desperation — and you must not expect a great deal from it.

Content Guidance

Module 3 Periodic table and energy

The periodic table

Periodicity

Periodicity is defined as the trends in physical and chemical properties that are repeated from one period to the next.

The periodic table is the arrangement of elements by increasing atomic number. Elements with the same outer-shell electron configuration are grouped together in periods showing repeating trends in physical and chemical properties.

Trends in the periodic table

Atomic radius

This decreases across a period, because the attraction between the nucleus and outer electrons increases. This is because across a period:

- the nuclear charge increases
- outer electrons are being added to the same shell, so there is no extra shielding

Atomic radius increases down a group because the attraction between the nucleus and outer electrons decreases. This is because down a group:

- extra shells are added, resulting in the outer shell being further from the nucleus
- more shells between the outer electrons and the nucleus mean greater shielding

Electrical conductivity, melting point and boiling point are related to structure and bonding, as shown in Table 1.

Type of structure	Giant lattice				Simple molecular		
Element	Na	Mg	Al	Si	P (P_4)	S (S_8)	Cl (Cl_2)
Forces	Strong forces throughout lattice				Weak dipole–dipole interactions between individual molecules		
Type of bonding	Metallic			Covalent			
Melting point	High				Low		
Conductivity	Good			Poor			

Table 1

Electrical conductivity

Elements in groups 1, 2 and 3 are metals. They are good conductors. Metallic elements are good conductors because they contain mobile, delocalised electrons. The outer-shell electrons can contribute to the mobile, delocalised electrons, allowing metals to conduct heat and electricity, even in the solid state.

The elements in the remaining groups across periods 2 and 3 are poor conductors because they do not have any mobile, delocalised electrons. Graphite and graphene are exceptions and are good conductors because they have mobile delocalised electrons.

Group 2 elements tend to be better conductors than group 1, because they have two outer-shell electrons, while group 1 elements only have one outer-shell electron.

Melting points and boiling points

These show a gradual increase from groups 1, 2, 3 and 14, followed by a sharp drop to groups 15, 16 and 17 (Figure 1).

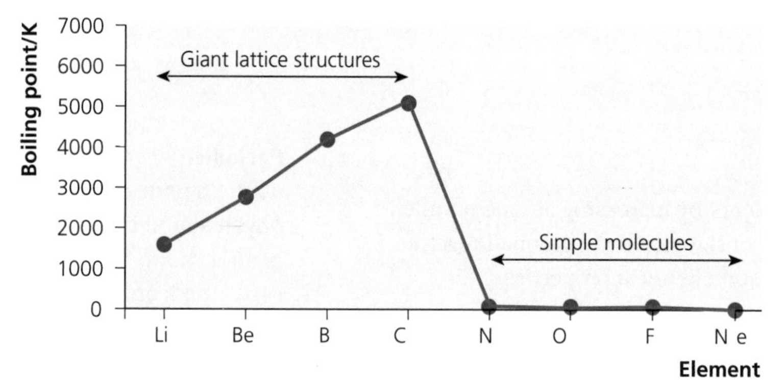

Figure 1 Trends in boiling point in period 2. This trend is repeated in period 3

The large drop signifies the move from giant structures in groups 1–14, to simple molecular structures in groups 15–17.

Ionisation energy

Ionisation energy decreases down a group because the outer electrons are further from the nucleus and are shielded by additional inner shells. This results in a decrease in the effective nuclear charge.

Ionisation energy increases across a period due to an increase in nuclear charge and a decrease in atomic radii. Within this general trend there are small 'dips' after group 2 and after group 15.

Exam tip

If asked to describe trends in ionisation energy it is best to focus on three factors (Table 2).

Factor 1	Atomic radius increases down a group	Therefore easier to lose an electron
Factor 2	Shielding increases down a group	Therefore easier to lose an electron
Factor 3	Nuclear charge increases down a group	Therefore harder to lose an electron

Table 2

You can use these trends to explain the reactivity down groups 2 and 17.

Knowledge check 1

The grid below shows the successive melting points of 4 elements in period 3.

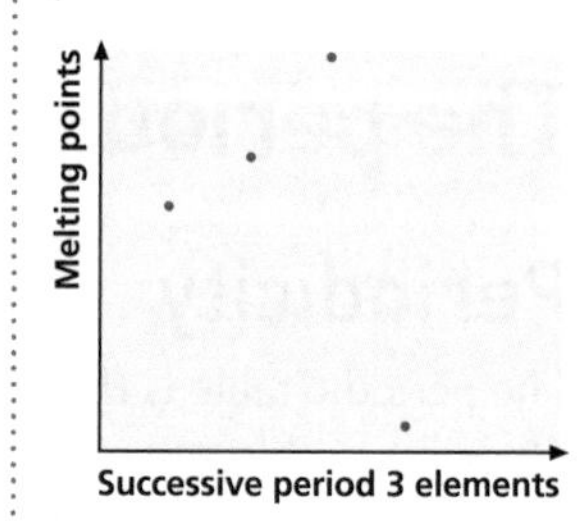

Are the elements:

A Na, Mg, Al, Si;
B Mg, Al, Si, P;
C Al, Si, S, P;
D Si, P, S, Cl?

Justify your answer.

Knowledge check 2

Put the elements Na, K, Mg, Al in order of increasing:

a atomic radii
b conductivity
c ionisation energies

Group 2

Properties of group 2 elements

Electronic configuration

Each group 2 element has two electrons in its outer shell and readily forms a 2+ ion, which has the same electron configuration as a noble gas as illustrated in Table 3 by magnesium and calcium.

Electron configuration of group 2 atom	Electron configuration of group 2 ion
$_{12}Mg$ $1s^2\,2s^2\,2p^6\,3s^2$	$_{12}Mg^{2+}$ $1s^2\,2s^2\,2p^6$
$_{20}Ca$ $1s^2\,2s^2\,2p^6\,3s^2\,3p^6\,4s^2$	$_{20}Ca^{2+}$ $1s^2\,2s^2\,2p^6\,3s^2\,3p^6$

Table 3

Physical properties

Group 2 elements are metallic and are therefore good conductors. They generally form colourless ionic compounds that tend to:

- have reasonably high melting and boiling points
- react with water
- be good conductors when molten or aqueous, but poor conductors when solid

Redox reactions of group 2 elements

You should be able to use oxidation numbers to illustrate the redox reactions that occur when group 2 elements react with oxygen and with water.

Reactions such as $2Mg(s) + O_2(g) \rightarrow 2MgO(s)$ are known as redox reactions.

Oxidation and **reduction** can be defined in terms of electrons or in terms of oxidation number.

Oxidation involves the loss of electrons or an increase in the oxidation number.

Reduction involves the gain of electrons or a decrease in the oxidation number.

These definitions can both be applied to the reaction in Figure 2.

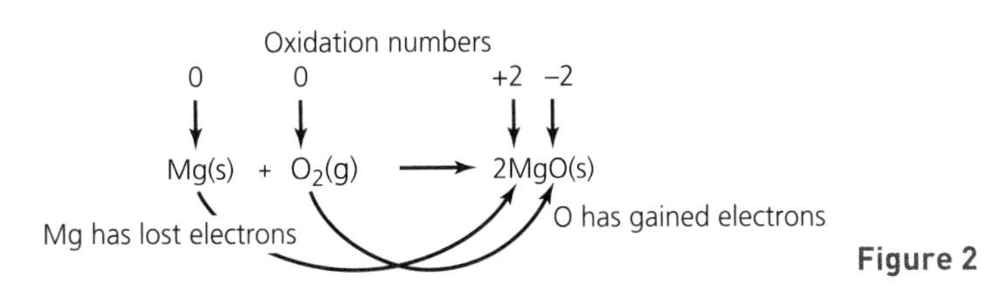

Figure 2

Reaction of group 2 elements with oxygen

Calcium, strontium and barium also react with oxygen to produce their oxides, but reactivity increases down the group. This is explained by the increasing ease with which the group 2 elements form the corresponding 2+ ion.

These are redox reactions, in that the oxidation number of the group 2 element increases from 0 to +2, while the oxidation number of oxygen decreases from 0 to −2.

Reaction of group 2 elements with water

Group 2 elements also undergo a redox reaction with water:

$$M(s) + 2H_2O(l) \rightarrow M(OH)_2(aq) + H_2(g)$$

where M is Mg, Ca, Sr or Ba. $Mg(OH)_2(s)$ is sparingly soluble and the state symbol is best shown as (s).

Each of these reactions is also a redox reaction. Once again the oxidation number of the group 2 element increases from 0 to +2, but this time the oxidation number of hydrogen changes from +1 to 0.

The reaction between magnesium and water is very slow, and the resultant $Mg(OH)_2$ is barely soluble in water, forming a white suspension.

Magnesium also reacts with steam, to produce magnesium oxide and hydrogen:

$$Mg(s) + H_2O(g) \rightarrow MgO(s) + H_2(g)$$

The rate of reaction increases down the group, largely due to the ease of cation (M^{2+}) formation.

Reaction of group 2 oxides with water

All group 2 metal oxides react with water to form hydroxides:

$$MO(s) + H_2O(l) \rightarrow M(OH)_2(aq)$$

where M is Mg, Ca, Sr or Ba. $Mg(OH)_2(s)$ is a suspension and $Ca(OH)_2(aq)$ is limewater.

These are *not* redox reactions. The oxidation numbers of all the elements remain the same.

The resulting hydroxide solutions are alkaline and have pH values of 8–12.

$$M(OH)_2(s) \rightleftharpoons M^{2+}(aq) + 2OH^-(aq)$$

Alkalinity depends on the concentration of hydroxide ions in solution, $[OH^-(aq)]$. $Mg(OH)_2$ is almost insoluble so the concentration of hydroxide ions in solution is low, as is the alkalinity. Solubility of the group 2 hydroxides increases down the group, as does alkalinity.

Calcium hydroxide is used in agriculture to neutralise acidic soils, while magnesium hydroxide is used in some indigestion tablets as an antacid.

Knowledge check 3

Group 2 elements react with water. Write an equation for the reaction when Sr reacts with water. Explain the trend in reactivity from Mg to Ba.

The halogens (group 17)

Electronic configuration

Each group 17 element has seven electrons in its outer shell and readily forms a 1– ion (an anion) with the same electronic configuration as a noble gas (Table 4).

Electron configuration of group 17 atom, X	Electron configuration of group 17 ion, X^-
$_9F$ $1s^2\,2s^2\,2p^5$	$_9F^-$ $1s^2\,2s^2\,2p^6$
$_{17}Cl$ $1s^2\,2s^2\,2p^6\,3s^2\,3p^5$	$_{17}Cl^-$ $1s^2\,2s^2\,2p^6\,3s^2\,3p^6$
$_{35}Br$ $1s^2\,2s^2\,2p^6\,3s^2\,3p^6\,3d^{10}\,4s^2\,4p^5$	$_{35}Br^-$ $1s^2\,2s^2\,2p^6\,3s^2\,3p^6\,3d^{10}\,4s^2\,4p^6$

Table 4

Physical properties

Element		State	Colour	Volatility
Fluorine	F_2	Gas	Yellow	As you go down the group there is an increase in dipole–dipole interactions (van der Waals forces) corresponding to an increased number of electrons in the halogen molecules. This increase reduces the volatility and hence raises the melting and boiling points
Chlorine	Cl_2	Gas	Green	
Bromine	Br_2	Liquid	Orange/brown	
Iodine	I_2	Solid	Brown/black	
The halogens are all non-metallic, making them poor conductors of electricity				

Table 5

Chemical properties

The reactivity of the group 17 halogens decreases down the group (opposite to the trend for the group 2 metals, which react by losing electrons). The halogens react not by losing electrons but by gaining an electron to form a halide anion. The ease with which the electron is gained decreases down the group. This is because atomic radius and shielding both increase down the group, reducing the effective attraction of the nucleus for electrons. Fluorine is the most reactive of the halogens and iodine is the least reactive. Fluorine is a powerful oxidising agent and readily gains electrons.

Knowledge check 4

Explain why fluorine is more reactive than chlorine.

It is important to know the difference between a halogen and a halide. In examinations, many students confuse chloride with chlorine. A halogen (fluorine, chlorine and bromine) will displace a halide (Cl^-, Br^- and I^-) from one of its salts. This is shown clearly in Table 6.

	Fluoride, F^-	Chloride, Cl^-	Bromide, Br^-	Iodide, I^-
Fluorine, F_2		✓	✓	✓
Chlorine, Cl_2	✗		✓	✓
Bromine, Br_2	✗	✗		✓
Iodine, I_2	✗	✗	✗	

Table 6

Exam tip

Students get very confused between the halogens and the halides. Look carefully at the name: if it ends ...ine it tells you that it is a diatomic element, but if it ends ...ide this indicates that it is a negative ion, e.g. chlorine is Cl_2 while chloride is Cl^-.

Displacement reactions of the halogens

- Fluorine displaces chloride, bromide and iodide ions from solution.
- Chlorine displaces bromide and iodide ions.
- Bromine only displaces iodide ions.
- Iodine does not displace any of the halides above.

This trend illustrates the decrease in oxidising power down the group.

Chlorine oxidises both bromide and iodide ions:

$$Cl_2(aq) + 2Br^-(aq) \rightarrow 2Cl^- + Br_2(l)$$

During the above reaction the formation of the orange/brown colour of bromine can be observed. If an organic solvent is added, the orange/brown colour of the bromine will intensify in the upper organic layer.

$$Cl_2(aq) + 2I^-(aq) \rightarrow 2Cl^- + I_2(s)$$

Bromine oxidises iodide ions only:

$$Br_2(aq) + 2I^-(aq) \rightarrow 2Br^- + I_2(s)$$

During the above reaction the brown/black colour of iodine can be seen. If an organic solvent is added, a distinctive violet colour in the upper organic layer is seen.

Iodine does *not* oxidise either chloride or bromide ions.

The displacement reactions are redox reactions. In each case, the halogen higher in the group gains electrons (is reduced) to form the corresponding halide (Figure 3).

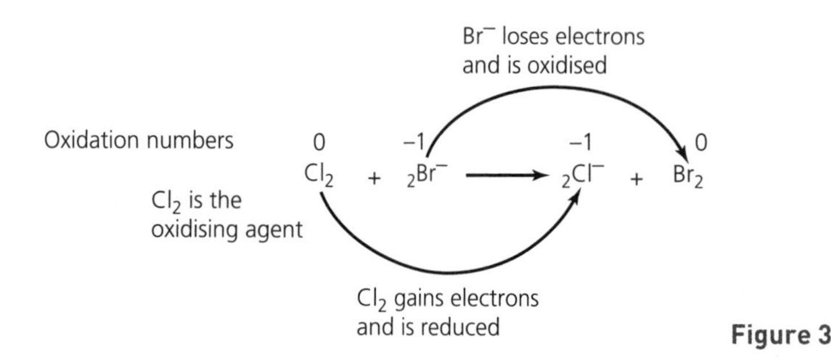

Figure 3

Testing for halide ions

Silver chloride, bromide and iodide are insoluble in water and therefore the chloride, bromide and iodide can be detected by the addition of a solution of silver nitrate ($AgNO_3(aq)$). Each of the silver halides forms a different-coloured precipitate. The precipitates can be distinguished by their solubility in ammonia.

AgCl is a white precipitate, which is soluble in dilute ammonia.

$$Ag^+(aq) + Cl^-(aq) \rightarrow AgCl(s)$$

AgBr is a cream precipitate, which is soluble in concentrated ammonia.

$$Ag^+(aq) + Br^-(aq) \rightarrow AgBr(s)$$

AgI is a yellow precipitate, which is insoluble in concentrated ammonia.

$$Ag^+(aq) + I^-(aq) \rightarrow AgI(s)$$

Uses of chlorine

Chlorine is used in water treatment. The gas reacts with water in a reversible reaction, and the resultant mixture kills bacteria, making the water safe to drink. However, chlorine can also react with hydrocarbons in the water, forming chlorinated hydrocarbons, which present a health risk.

$$Cl_2 + H_2O \rightleftharpoons HCl + HClO$$

The reaction of chlorine with water is a redox reaction, but is unusual in the sense that the chlorine atom undergoes both oxidation and reduction. This is known as **disproportionation.**

One chlorine atom in the Cl_2 molecule is oxidised as the oxidation number changes from 0 to +1, while the other is reduced from 0 to −1.

Disproportionation is the simultaneous oxidation and reduction of an element such that during a reaction its oxidation number both increases and decreases.

Under certain conditions chlorine can also react with NaOH to form bleach. In this reaction the chlorine also undergoes disproportionation.

$Cl_2 + 2NaOH \rightarrow NaCl + NaClO + H_2O$ NaClO is sodium chlorate(I)

If the conditions are changed, disproportionation stills occurs but it is possible to form different chlorate salts:

$2Cl_2 + 4NaOH \rightarrow 3NaCl + NaClO_2 + 2H_2O$ $NaClO_2$ is sodium chlorate(III)

$3Cl_2 + 6NaOH \rightarrow 5NaCl + NaClO_3 + 3H_2O$ $NaClO_3$ is sodium chlorate(V)

Qualitative analysis

Test tube reactions for anions and for cations

You will be expected to analyse and detect a range of ions by a series of test tube reactions. Table 7 details qualitative tests for a range of ions.

ion	test	equation	observation
CO_3^{2-}	Add an acid, $H^+(aq)$	$CO_3^{2-}(aq) + 2H^+(aq) \rightarrow CO_2(g) + H_2O(l)$	Effervescence, bubbles
SO_4^{2-}	Add aqueous $BaCl_2(aq)$	$SO_4^{2-}(aq) + Ba^{2+}(aq) \rightarrow BaSO_4(s)$	White precipitate
Cl^-	Add $AgNO_3(aq)$	$Cl^-(aq) + Ag^+(aq) \rightarrow AgCl(s)$	White precipitate *
Br^-		$Br^-(aq) + Ag^+(aq) \rightarrow AgBr(s)$	Cream precipitate *
I^-		$I^-(aq) + Ag^+(aq) \rightarrow AgI(s)$	Yellow precipitate *
NH_4^+	Warm with NaOH(aq)	$NH_4^+ + OH^-(aq) \rightarrow NH_3(g) + H_2O(l)$	$NH_3(g)$ is evolved which turns moist red litmus blue

* The silver halide precipitates can be further distinguished by adding ammonia to the precipitates. See p. 12.

Table 7

Summary

Having revised **Module 3: The periodic table** you should now have an understanding of:

- trends in atomic radii, melting points/boiling points and ionisation energies
- reactions of group 2 elements with oxygen and with water
- reactions of group 2 oxides with water
- decomposition of group 2 carbonates
- physical properties of the halogens
- displacement reactions of the halogens and the halides
- reactions of halides with Ag^+ and NH_3
- disproportionation reactions of chlorine

Physical chemistry

Enthalpy changes

ΔH of reaction, formation and combustion

Enthalpy change is the exchange of energy between a reaction mixture and its surroundings and is given the symbol ΔH. The units are always kJ mol^{-1}.

If the reaction mixture loses energy to its surroundings, then the reaction is **exothermic** and ΔH is *negative*.

If the reaction mixture gains energy from its surroundings, the reaction is **endothermic** and ΔH is *positive*.

ΔH can be calculated using the equation:

ΔH = enthalpy of products – enthalpy of reactants

Enthalpy changes can be represented by simple enthalpy-profile diagrams.

For an exothermic reaction, the enthalpy-profile diagram shows the enthalpy of products at a lower energy than the enthalpy of reactants. For an endothermic reaction, the enthalpy-profile diagram shows the enthalpy of products at a higher energy than the enthalpy of reactants. The difference in the enthalpy is ΔH. E_a is the activation energy (Figure 4).

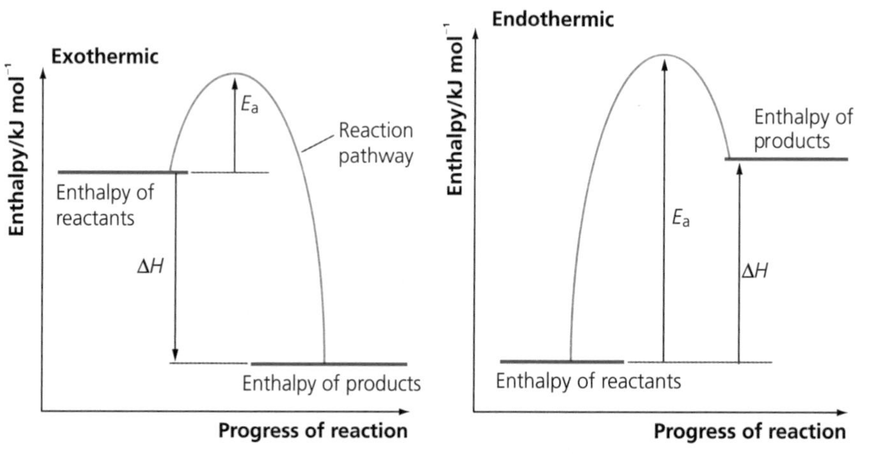

Figure 4

Oxidation reactions such as the combustion of fuels are exothermic and release energy to their surroundings. This results in an increase in temperature in the surroundings.

Oxidation of carbohydrates, such as glucose, in respiration is also an exothermic reaction.

Thermal decomposition reactions are usually endothermic, requiring energy from the surroundings.

Enthalpy-profile diagrams not only show the enthalpy change of reaction, ΔH, but also display the activation energy, E_a.

Exam tip

The symbol 'Δ' indicates a change:

- ΔT = change in temperature
- ΔV = change in volume
- ΔP = change in pressure

Activation energy is defined as the minimum energy required, in a collision between particles, if they are to react. (This is covered in more detail in the section on reaction rates; see p. 21.) In any chemical reaction, bonds have to be broken and new bonds have to be formed. Breaking bonds is an endothermic process, requiring energy. This energy requirement contributes to the activation energy of a reaction.

Knowledge check 5

Define the term activation energy, E_a.

Knowledge check 6

Draw a fully labelled energy-profile diagram for:

a $2SO_2(g) + O_2(g) \rightarrow 2SO_3(g)$ $\Delta H = -196\,kJ\,mol^{-1}$

b $2H_2(g) + 2I_2(g) \rightarrow 2HI(g)$ $\Delta H = +53\,kJ\,mol^{-1}$

Standard enthalpy changes

All standard enthalpy changes are measured under standard conditions. The temperature and the pressure at which measurements and/or calculations are carried out are standardised.

Standard temperature = 298 K (25°C)

Standard pressure = 101 kPa ($100\,000\,N\,m^{-2} = 10^5\,Pa$ = 1 bar = 1 atmosphere)

(For examination purposes, 100 kPa is acceptable.)

Standard temperature and pressure are often referred to as STP.

Examinations often ask for a definition of enthalpy changes, and it is advisable to learn the following definitions.

Standard enthalpy change of reaction

$\Delta_r H^\ominus$ is the enthalpy change when the numbers of moles of the substances in the balanced equation react under the standard conditions of 298 K and 100 kPa.

$2SO_2(g) + O_2(g) \rightarrow 2SO_3(g)$

Standard enthalpy change of formation

$\Delta_f H^\ominus$ is the enthalpy change when one mole of a substance is formed from its elements, in their natural state, under the standard conditions of 298 K and 100 kPa.

$S(g) + 1\frac{1}{2}O_2(g) \rightarrow SO_3(g)$

Standard enthalpy change of combustion

$\Delta_c H^\ominus$ is the enthalpy change when one mole of a substance is burnt completely, in an excess of oxygen, under the standard conditions of 298 K and 100 kPa.

$CH_4(g) + 2O_2(g) \rightarrow CO_2(g) + 2H_2O(l)$

Standard enthalpy change of neutralisation

$\Delta_{neut} H^\ominus$ is the enthalpy change when one mole of water is formed from neutralisation (by reacting an acid with a base), under the standard conditions of 298 K and 100 kPa.

$HNO_3(aq) + LiOH(aq) \rightarrow LiNO_3(aq) + H_2O(l)$

Bond enthalpies

Average bond enthalpy

This is the enthalpy change on breaking one mole of a covalent bond in a gaseous molecule under the standard conditions of 298 K and 100 kPa.

In addition to these definitions, you may also be expected to show your understanding by writing equations to illustrate the standard enthalpy changes of reaction, formation and combustion.

The equation for the standard enthalpy of neutralisation of HCl(aq) with NaOH(aq) is:

$$HCl(aq) + NaOH(aq) \rightarrow NaCl(aq) + H_2O(l)$$

An ionic equation can also be used to show the standard enthalpy of neutralisation:

$$H^+(aq) + OH^-(aq) \rightarrow H_2O(l)$$

Equations to show the standard enthalpy of formation of a substance must reflect the standard definition and must always:

- show the elements as the reactants
- produce *one* mole of the substance, even if that means having fractions in the equation
- show the state symbols

The equation for the standard enthalpy of formation of:

- ethane is $2C(s) + 3H_2(g) \rightarrow C_2H_6(g)$
- ethanol is $2C(s) + 3H_2(g) + ½O_2(g) \rightarrow CH_3CH_2OH(l)$

Equations to show the standard enthalpy of combustion of a substance must reflect the standard definition and must always:

- react *one* mole of the substance with excess $O_2(g)$, even if that means having fractions in the equation
- show the state symbols

Usually, the products are $CO_2(g)$ and $H_2O(l)$.

The equation for the standard enthalpy of combustion of:

- ethane is $C_2H_6(g) + 3½O_2 \rightarrow 2CO_2(g) + 3H_2O(l)$
- ethanol is $CH_3CH_2OH + 3O_2 \rightarrow 2CO_2(g) + 3H_2O(l)$

You will be expected to calculate enthalpy changes:

- using enthalpy data from experimental data
- average bond enthalpies
- by using Hess's law

Knowledge check 7

Define standard enthalpy change of formation and write an equation to illustrate the standard enthalpy change of formation of propanone, $CH_3COCH_3(l)$.

Knowledge check 8

Define standard enthalpy change of combustion and write an equation to illustrate the standard enthalpy change of combustion of propanal, $CH_3CH_2CHO(l)$.

Enthalpy changes using enthalpy data from experiments

The standard enthalpy change, $\Delta_r H^\ominus$, for reactions that take place in solution can usually be measured directly by using the simple apparatus shown in Figure 5. However, the results obtained will only be approximate, since there are likely to be substantial heat losses to the surroundings.

The energy transfer for the reaction mixture is given the symbol q and can be calculated by using the equation:

$$q = mc\Delta T$$

where m is the mass of the reaction mixture, c is the specific heat capacity of the reaction mixture and ΔT is the change in temperature.

The solvent used for most reactions is water. The specific heat capacity of water, c, is taken as either $4.2\,\text{J}\,\text{g}^{-1}\,\text{K}^{-1}$ or $4.2\,\text{kJ}\,\text{kg}^{-1}\,\text{K}^{-1}$. The enthalpy change for the reaction mixture will give a value in either J (joules) or kJ (kilojoules), depending on the value/units of the specific heat capacity. It is usual to correct this value (q), so that the ΔH value can be quoted for one mole of reactant and the units of ΔH become $\text{kJ}\,\text{mol}^{-1}$. The standard enthalpy change for the reaction can then be calculated by dividing the energy transferred by the number of moles, n, of reactant used.

$$\Delta_r H = \frac{q}{n} = \frac{mc\Delta T}{n}$$

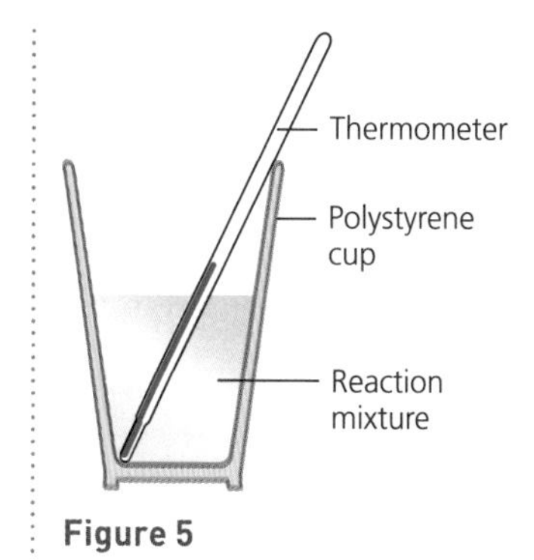

Figure 5

Exam tip

In a question such as:

2.30 g ethanol was burnt and raised the temperature of 300 cm³ water by 12.5°C. Calculate $\Delta_c H$ for ethanol.

you would be told the density of the water as well as the specific heat capacity, c, of the apparatus. Many make an error in the first step when using $q = mc\Delta T$ and incorrectly use 2.30 g as the mass. The mass should be 300 g, which is the mass of the water.

Example

When $50.0\,\text{cm}^3$ of $2.00\,\text{mol}\,\text{dm}^{-3}$ hydrochloric acid was mixed with $50.0\,\text{cm}^3$ of $2.00\,\text{mol}\,\text{dm}^{-3}$ sodium hydroxide, the temperature increased by 13.7°C. Calculate the standard enthalpy change of neutralisation of HCl(aq). Assume that the specific heat capacity, c, is $4.20\,\text{J}\,\text{g}^{-1}\,\text{K}^{-1}$ and that the densities of the HCl and NaOH are both $1.00\,\text{g}\,\text{cm}^{-3}$.

Method

Step 1: calculate the energy transferred in the reaction by using the equations

$q = mc\Delta T$

m = the mass of the two solutions = 50 + 50 = 100 g

$q = mc\Delta T = (100) \times (4.20) \times (13.7) = 5754\,\text{J} = 5.754\,\text{kJ}$

Step 2: convert the answer to $\text{kJ}\,\text{mol}^{-1}$ by dividing by the number of moles used

amount in moles of HCl, $n = cV = (2.00) \times (50/1000) = 0.1$ mol

$\Delta H = q/n = 5.754/0.1 = 57.54 = 57.5\,\text{kJ}\,\text{mol}^{-1}$

Remember that, because the temperature increased, it is an exothermic reaction and therefore $\Delta H = -57.5\,\text{kJ}\,\text{mol}^{-1}$.

The standard enthalpy change of combustion, $\Delta_c H^{\ominus}$, for a volatile liquid can also be measured directly, but again heat losses mean that the result is only approximate.

Knowledge check 9

2.48 g ethane-1,2-diol, $HOCH_2CH_2OH$, was burnt and raised the temperature of 500 cm³ water by 22.5°C.

Calculate $\Delta_c H$ for ethane-1,2-diol. The specific heat capacity, c, of the apparatus is $4.18\,\text{J}\,\text{g}^{-1}\,\text{K}^{-1}$

Enthalpy changes using average bond enthalpy data

Breaking a bond *requires energy*, while making a bond *releases energy*. The energy required to break a bond is exactly the same as the energy released when the same bond is made.

The bond enthalpy may be defined as the enthalpy change required to break and separate one mole of bonds in the molecules of a gas, so that the resulting gaseous (neutral) particles, atoms or radicals exert no forces upon each other. It is best reinforced by a simple equation such as:

$$Cl{-}Cl(g) \rightarrow Cl\bullet(g) + Cl\bullet(g)$$

or generally:

$$X{-}Y(g) \rightarrow X\bullet(g) + Y\bullet(g)$$

Bond enthalpies are the average (mean) values, and do not take into account the specific chemical environment.

Some average bond enthalpies are shown in Table 8.

Bond	ΔH/kJ mol^{-1}
C–H	+413
C=O	+805
O=O	+498
O–H	+464
C–N	+286
C=C	+612
N≡N	+945
N–H	+391
C–C	+347
H–H	+436

Table 8

Calculations involving average bond enthalpy

These calculations are straightforward. In order for a reaction to take place, existing bonds have to be broken (this needs an input of energy, i.e. it is endothermic), and then new bonds have to be formed (energy is given out, so this is exothermic). For a simple reaction involving gaseous molecules:

$$\Delta H = \Sigma \text{ bond enthalpies of reactants} - \Sigma \text{ bond enthalpies of products}$$

The enthalpy of combustion of methane can be calculated using average bond enthalpies, as shown below:

$$CH_4 + 2O_2 \rightarrow CO_2 + 2H_2O$$

It is useful to draw out the reaction using displayed formulae, so that all the bonds that are broken and formed can be seen clearly (Table 9).

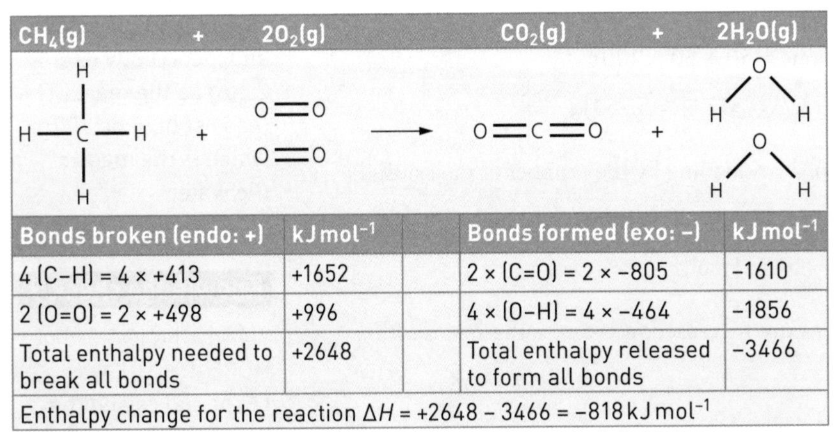

CH_4(g) + $2O_2$(g) → CO_2(g) + $2H_2O$(g)			
Bonds broken (endo: +)	**kJ mol^{-1}**	**Bonds formed (exo: –)**	**kJ mol^{-1}**
4 (C–H) = 4 × +413	+1652	2 × (C=O) = 2 × –805	–1610
2 (O=O) = 2 × +498	+996	4 × (O–H) = 4 × –464	–1856
Total enthalpy needed to break all bonds	+2648	Total enthalpy released to form all bonds	–3466
Enthalpy change for the reaction ΔH = +2648 – 3466 = –818 kJ mol^{-1}			

Table 9

The accepted value for this reaction is –890 kJ mol^{-1}, which differs substantially from the value calculated above. This can be explained largely by the fact that we use average bond enthalpies for the C–H, C=O and O–H bonds, and not specific bond enthalpies.

Knowledge check 10

Use the bond energies in Table 9 to calculate the enthalpy of hydrogenation of ethane, shown below.

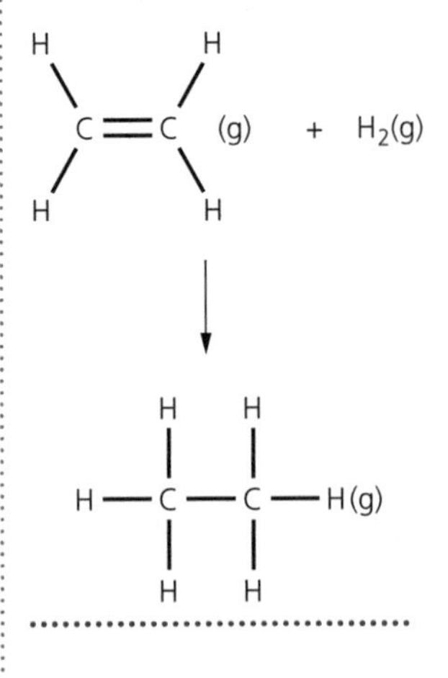

Enthalpy changes using Hess's law

Many chemical reactions cannot be measured directly. This can be for either energetic or kinetic reasons. If the activation energy is very high or if a reaction rate is very slow, then the enthalpy change cannot be measured directly by experiment. Reactions that cannot be measured directly can be calculated indirectly by using energy cycles. Energy cannot be created or destroyed and therefore the enthalpy change for a reaction is independent of the route taken.

Hess's law states that if a reaction can take place by more than one route, the overall enthalpy change for each route is the same, irrespective of the route taken, provided that the initial and final conditions are the same.

When applying Hess's law it is helpful to construct an enthalpy triangle.

It is difficult to measure the enthalpy of formation of CO(g) directly, but it can be calculated using Hess's law.

The enthalpy change for the following reactions can be measured experimentally:

$C(s) + O_2(g) \rightarrow CO_2(g) \qquad \Delta H = -394\,kJ\,mol^{-1}$

$CO(g) + ½O_2(g) \rightarrow CO_2(g) \qquad \Delta H = -284\,kJ\,mol^{-1}$

It is always best to start with the enthalpy change that has to be calculated. You can then write that along the top and call it ΔH_1.

$C(s) + ½O_2(g) \rightarrow CO(g)$

Then construct a cycle with two alternative routes (Figure 6).

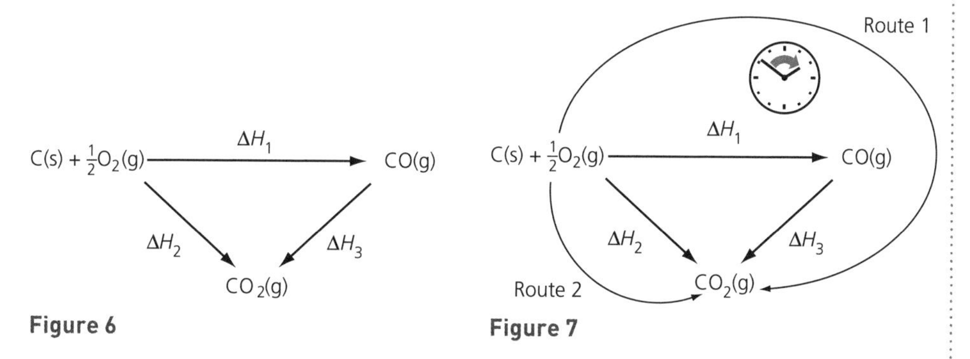

Figure 6 **Figure 7**

The simplest way to apply Hess's law is then to look at the direction of the arrows. Route 1 is made up of arrows that point in the clockwise direction, while route 2 consists of arrows that point in the anti-clockwise direction (Figure 7).

route 1 = route 2, therefore $\Delta H_1 + \Delta H_3 = \Delta H_2$

$\Delta H_1 = \Delta H_2 - \Delta H_3$

$= -394 - (-284)$

$= -394 + 284$

$= -110\,kJ\,mol^{-1}$

So, for $C(s) + ½O_2(g) \rightarrow CO(g)$, $\Delta H = -110\,kJ\,mol^{-1}$.

If asked to calculate an enthalpy change using Hess's law you will be supplied with data that will enable you to carry out the calculation.

- If the data you are supplied with is the enthalpy of combustion, it is best to construct a cycle like the one shown in Figure 8, where the combustion products are shown at the bottom. The arrows always point downwards to the combustion products.

If Hess's law is applied to the cycle:

$\Delta H_1 + \Delta H_3 = \Delta H_2$ hence $\Delta H_1 = \Delta H_2 - \Delta H_3$

- If the data you are supplied with is the enthalpy of formation, it is best to construct a cycle like the one shown in Figure 9, where the elements are shown at the bottom and the arrows always point up to the reactants and to the products.

If Hess's law is applied to the cycle:

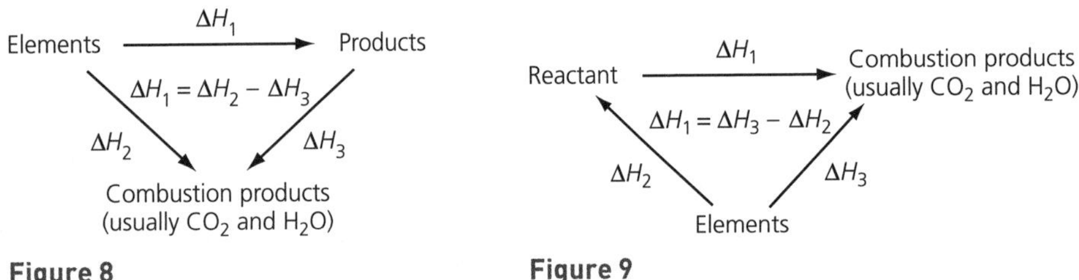

Figure 8 **Figure 9**

$\Delta H_1 + \Delta H_2 = \Delta H_3$ hence $\Delta H_1 = \Delta H_3 - \Delta H_2$

Example

Use the data in Table 10 to calculate the standard enthalpy change of combustion for ethane:

Substance	$\Delta_f H^{\ominus}$/kJ mol^{-1}
$C_2H_6(g)$	−85
$CO_2(g)$	−394
$H_2O(l)$	−286

Table 10

Step 1: write an equation for the enthalpy of combustion required:

$$C_2H_6(g) + 3\tfrac{1}{2}O_2(g) \rightarrow 2CO_2(g) + 3H_2O(l)$$

Step 2: construct the enthalpy triangle by writing the elements at the bottom. Both arrows will point upwards (Figure 10).

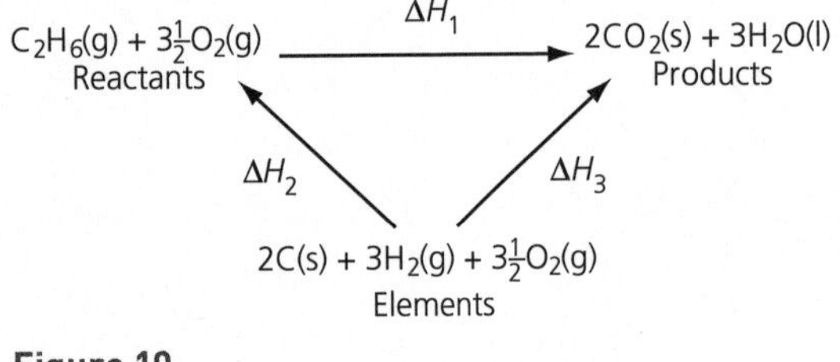

Figure 10

Exam tip

Most exam papers contain at least one question using Hess's law and there are only two types of question that can be asked. If the data provided are standard enthalpy of:

- combustion, the cycle will have combustion products at the bottom and the arrows will point down
- formation, the cycle will have elements at the bottom and the arrows will point up

Step 3: apply Hess's law using the clockwise = anti-clockwise rule:

$\Delta H_1 + \Delta H_2 = \Delta H_3$ hence $\Delta H_1 = \Delta H_3 - \Delta H_2$

ΔH_3 = formation of $C_2H_6(g) = -85\,kJ\,mol^{-1}$

$\Delta H_2 = 2 \times$ [formation of $CO_2(g)$] + $3 \times$ [formation of $H_2O(l)$]

$= (2 \times -394) + (3 \times -286)$

$= -1646\,kJ\,mol^{-1}$

therefore

$\Delta H_1 = (-1646) - (-85) = -1561\,kJ\,mol^{-1}$

$C_2H_6(g) + 3\tfrac{1}{2}O_2(g) \rightarrow 2CO_2(g) + 3H_2O(l) \quad \Delta_c H^{\ominus} = -1561\,kJ\,mol^{-1}$

Knowledge check 11

Use the data below to calculate the enthalpy of reaction for

$CuSO_4(s) + 5H_2O(l) \rightarrow CuSO_4.5H_2O(s)$

The enthalpies of formation, $\Delta_f H^{\ominus}$ are: $CuSO_4(s) = -771.4\,kJ\,mol^{-1}$, $H_2O(l) = -285.8\,kJ\,mol^{-1}$, $CuSO_4.5H_2O(s) = -2279.6\,kJ\,mol^{-1}$.

Reaction rates

Experimental observations show that the rate of a reaction is influenced by:

- temperature
- concentration
- use of a catalyst

The collision theory of reactivity helps to provide explanations for these observations. A reaction cannot take place unless a collision occurs between the reacting particles. Increasing the temperature or concentration increases the chance that a collision will occur.

However, not all collisions lead to a successful reaction. The energy of a collision between reacting particles must exceed the minimum energy required to start the reaction. This minimum energy is known as the activation energy, E_a. Increasing the temperature affects the number of collisions that have an energy exceeding E_a, while the use of a catalyst lowers E_a.

Knowledge check 12

Explain how increasing the pressure on a gaseous reaction affects the rate of reaction.

Boltzmann distribution of molecular energies

Energy is directly proportional to absolute temperature. When collisions occur, the particles involved in the collision exchange (gain or lose) energy, even if a reaction does not occur. It follows that for any given mass of gaseous reactants, at a constant temperature, the distribution of energies will mean that some particles have more energy than others.

Figure 11 shows a typical distribution of energies at constant temperature, known as the Boltzmann distribution.

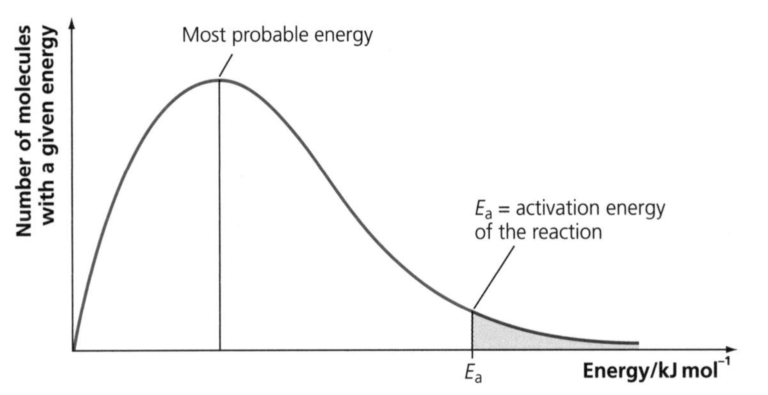

Figure 11

- The distribution always goes through the origin, showing that there are no particles with zero energy.
- The distribution is asymptotic (i.e. the curve approaches the axis, but will only meet the axis at infinity) to the horizontal axis at high energy. This shows that there is no maximum energy.
- E_a represents the activation energy, which is the minimum energy required to start the reaction.
- The area under the curve represents the total number of particles.
- The shaded area represents the number of particles with energy greater than or equal to the activation energy, $E \geq E_a$ (showing the number of particles with sufficient energy to react — Figure 12).

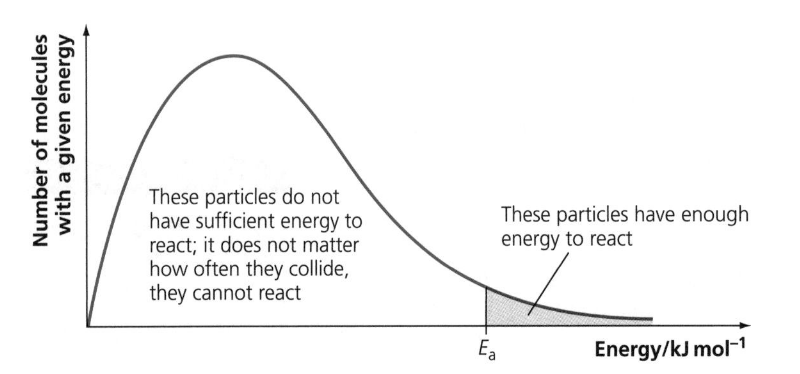

Figure 12

Effect of concentration, temperature and catalysts on the rate of reaction

Concentration

Increasing concentration simply increases the chance of a collision. The more collisions there are, the faster the reaction will be.

For a gaseous reaction, increasing the pressure has the same effect as increasing the concentration. If the pressure is increased, the volume will decrease causing the concentration to increase. When gases react, they react faster at high pressure, because as the pressure increases so does the concentration, leading to an increased chance of a collision.

Temperature

An increase in temperature has a dramatic effect on the distribution of energies, as can be seen in Figure 13.

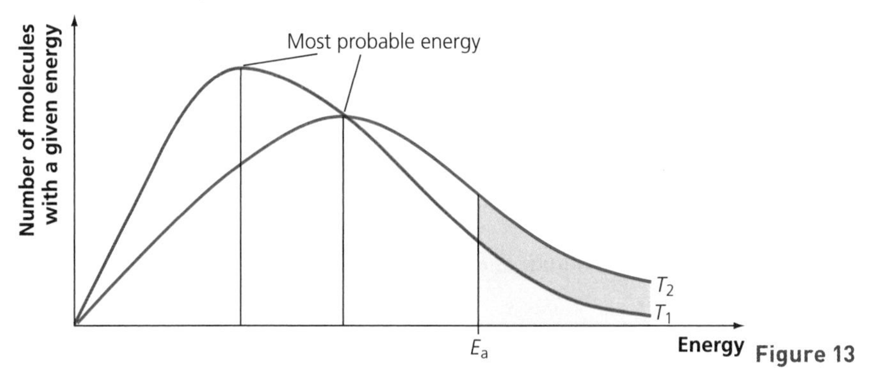

Figure 13

Knowledge check 13

Copy the graph shown in Figure 12. Label it as the distribution of molecular energies at 40°C. Add a second curve to represent the distribution of molecular energies at 25°C.

Only the temperature has changed. The number of particles remains the same, and therefore the area under both curves remains the same.

At higher temperature (T_2) the distribution flattens and shifts to the right, so that:

- there are fewer particles with low energy
- the most probable energy moves to higher energy
- more particles have $E \geq E_a$

Increasing temperature increases the number of particles with energies greater than or equal to the activation energy, $E \geq E_a$, so at high temperature there are more particles with sufficient energy to react, making the reaction faster.

Decreasing the temperature has the opposite effect.

Catalysts

From GCSE, you will recall that catalysts speed up reactions without themselves changing permanently. Catalysts work by lowering the activation energy for the reaction. This is illustrated by an energy-profile diagram (Figure 14) and by the Boltzmann distribution (Figure 11).

E_a is the activation energy of the uncatalysed reaction.
E_{cat} is the activation energy of the catalysed reaction.
E_{cat} is lower than E_a.

A catalyst lowers the activation energy but does not alter the Boltzmann distribution. It follows that more particles have $E \geq E_{cat}$.

The shaded areas in Figure 15 indicate the proportion of particles whose energy exceeds the activation energies.

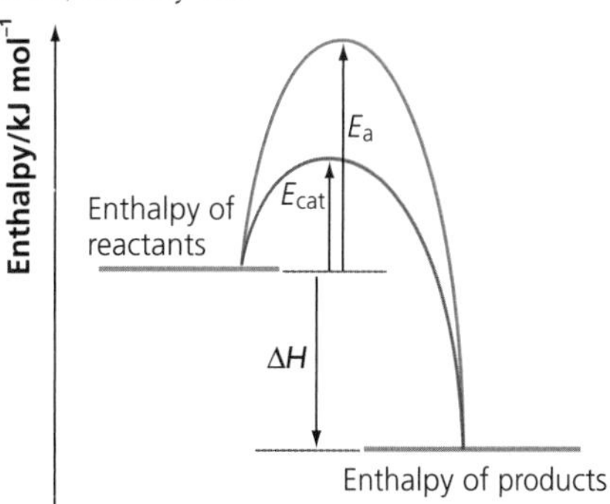

Figure 14

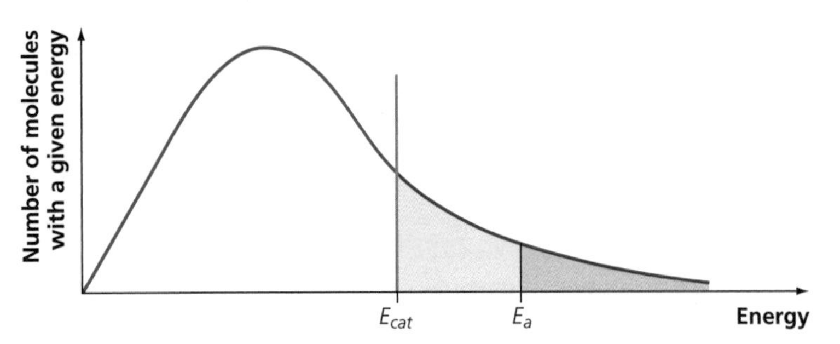

Figure 15

Homogeneous catalysts

The catalyst is in the same phase (gas, liquid or solid) as the reactants. The most common phase is a liquid, as many reactions are carried out in aqueous solution. A good example of a reaction involving a homogeneous catalyst is esterification, which you will meet in the second year of the A-level course.

The reaction involves ethanol (a liquid) and ethanoic acid (a liquid) and uses sulfuric acid (also a liquid) as the catalyst (Figure 16).

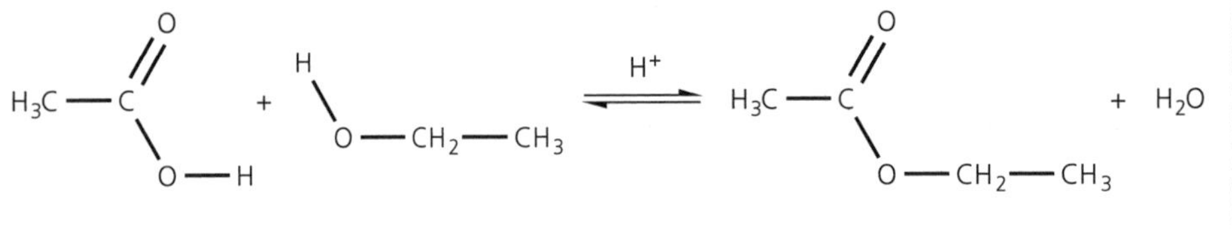

Figure 16

The acid catalyst works by providing an alternative mechanism involving an activated intermediate with a lower activation energy. The H^+ ions from the catalyst take part in the reaction, but they are reformed at the end. This means that the concentration of acid catalyst at the end of the reaction is the same as when the reagents were mixed.

A second example of a homogeneous catalyst is in the loss of ozone from the upper atmosphere (stratosphere). This is a gas-phase reaction and the catalyst is the chlorine radical, Cl•(g). The Cl•(g) is formed when the C–Cl bonds in CFCs are exposed to high-energy ultraviolet light, forming radicals. The reactions are complex, but can be summarised as shown on p. 49.

Heterogeneous catalysts

The catalyst is in a different phase to the reactants. The most common type of heterogeneous catalysis involves reactions of gases with a solid catalyst. Heterogeneous catalysts also lower the activation energy, but their mode of action is different to that of a homogeneous catalyst.

Heterogeneous catalysts work by adsorbing gases on to their solid surfaces. This adsorption results in a weakening of the bonds within the reactant molecules, thus lowering the activation energy. Bonds are broken and new bonds are formed. The product molecules are then desorbed from the solid surface of the catalyst.

Exam tip

When drawing the Boltzmann distribution at a higher temperature the marking points are:

- the curve goes through the origin and there are fewer particles with low energy
- the most probable energy moves to right (higher energy) but the height of the peak is lower
- there are more particles with high energy such that a greater proportion of particles have energy that exceeds the activation energy

Transition metals are frequently used as heterogeneous catalysts. Iron is used as the catalyst in the Haber process for producing ammonia. The iron is usually either finely divided (and therefore has a large surface area), or is porous and contains a small amount of metal oxide promoters.

$$N_2(g) + 3H_2(g) \xrightarrow[\text{as catalyst}]{Fe(s)} 2NH_3(g)$$

The motor car, with the internal combustion engine, is responsible for many of the pollutants discharged into the atmosphere. However, modern cars are now fitted with catalytic converters, which have contributed significantly to an improvement in air quality. Cars often discharge unreacted hydrocarbons into the atmosphere. Some of these, for example benzene, are toxic and carcinogenic.

Carbon monoxide is also formed by the incomplete combustion of fuel.

$$C_8H_{18} + 8\tfrac{1}{2}O_2 \rightarrow 8CO + 9H_2O$$

The internal combustion engine often reaches temperatures of around 1000°C. These high temperatures provide sufficient energy for nitrogen and oxygen (both present in the air) to react and form oxides of nitrogen.

$$N_2 + O_2 \rightarrow 2NO$$

$$2NO + O_2 \rightarrow 2NO_2$$

Carbon monoxide, oxides of nitrogen (NO and NO_2) and unburnt hydrocarbons can lead to the formation of photochemical smog. Under certain conditions, such as bright sunlight and still air conditions, this can produce low-level ozone. High-level ozone, which occurs in the upper atmosphere, is beneficial, but low-level ozone traps pollutant gases.

The catalytic converter, fitted to all modern cars, reduces the emission of unburnt hydrocarbons, carbon monoxide and oxides of nitrogen. The equations below show how this is achieved.

Removal of unburnt hydrocarbons:

$$C_8H_{18}(g) + 12\tfrac{1}{2}O_2(g) \rightarrow 8CO_2(g) + 9H_2O(g)$$

Removal of carbon monoxide and oxides of nitrogen:

$$2NO(g) + 2CO(g) \rightarrow N_2(g) + 2CO_2(g)$$

$$2NO_2(g) + 4CO(g) \rightarrow N_2(g) + 4CO_2(g)$$

All of the above reactions occur in the gas phase. The catalytic converter consists of a fine aluminium mesh coated with a thin, solid film of an alloy of platinum, rhodium and palladium. Catalytic converters only function efficiently at high temperatures and are therefore not very effective on short journeys.

Catalysts are of great economic importance. The use of a catalyst often allows:

- reactions to be carried out at lower temperatures and pressures, thus reducing costs
- different reactions to be used, with lower atom economy and therefore with reduced waste

Enzymes, which operate close to room temperature and pressure, are increasingly used to generate a specific product.

Exam tip

If you are asked how a heterogeneous catalyst works, look at how many marks are available. If there is more than 1 mark, the explanation required is usually that the catalyst works by first adsorbing gases onto its surface ✓, which weakens the bonds and lowers activation energy and leads to a chemical reaction ✓ followed by the products of the reaction desorbing from the surface of the catalyst ✓.

Chemical equilibrium

Dynamic equilibrium and le Chatelier's principle

Reversible reactions

There are many everyday examples of reversible reactions or processes — the most common being the physical states of water. If the temperature of water falls below 0°C, the water will freeze and ice will form. However, when the temperature rises above 0°C, the ice melts and reforms the water. This process can be represented as:

$$H_2O(s) \rightleftharpoons H_2O(l)$$

The $\rightleftharpoons$ sign indicates that a reaction is reversible. This type of reaction is common in many chemical reactions, such as esterification.

Ethanoic acid reacts with ethanol, in the presence of an acid catalyst, to produce ethyl ethanoate and water. However, the ester, ethyl ethanoate, is hydrolysed by water, in the presence of an acid catalyst, and reforms ethanoic acid and ethanol.

Dynamic equilibrium

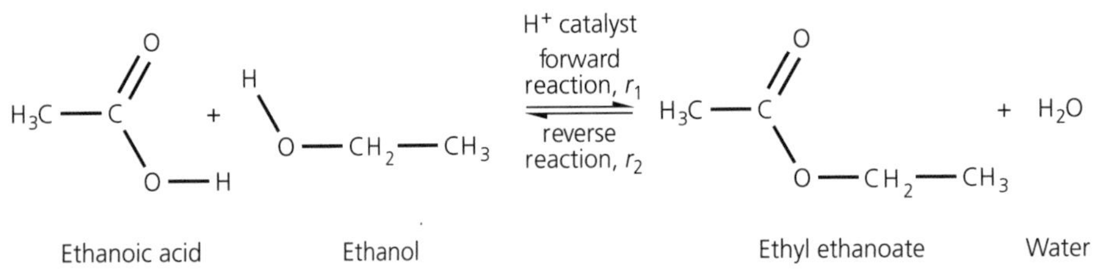

Figure 17

In the reversible reaction shown in Figure 17, the reaction moving from left to right is known as the **forward reaction**, and the reaction moving from right to left is referred to as the **reverse reaction**.

The forward reaction r_1 starts fast but slows down, while the reverse reaction r_2 starts slowly and speeds up. It follows that there will be a point at which the rate of the forward reaction exactly equals the rate of the reverse reaction: $r_1 = r_2$

When this happens, the system is said to be in **dynamic equilibrium**, i.e. equilibrium has been reached because the amount of each chemical in the system remains constant. It is dynamic because the reagents and the products are constantly interchanging. Dynamic equilibrium can only be reached if the system is closed.

Dynamic equilibrium is when the rate of the forward reaction equals the rate of the reverse reaction. The concentration of the reagents and products remains constant, while the reagents and the products constantly interchange.

le Chatelier's principle

The French chemist Henri le Chatelier studied many dynamic equilibria. He suggested a general qualitative rule to predict the movement of the position of the equilibrium.

le Chatelier's principle states that if a closed system under equilibrium is subject to a change, then the system will move in such a way as to minimise the effect of the change.

The factors that can be readily changed are concentration, temperature and pressure, and the use of a catalyst.

Effect of changing concentration on the position of the equilibrium

The equilibrium formed between the chromate ion, CrO_4^{2-}, and the dichromate ion, $Cr_2O_7^{2-}$, is useful because each ion is coloured, and it is therefore possible to observe the movement of the position of the equilibrium.

$$2H^+ + 2CrO_4^{2-} \rightleftharpoons Cr_2O_7^{2-} + H_2O$$

yellow orange

Knowledge check 14

State le Chatelier's principle and explain what is meant by the terms *reversible reaction* and *dynamic equilibrium*.

If we add some acid to the chromate/dichromate mixture, the concentration of the acid, $[H^+]$, increases. The system will now move in such a way as to minimise the effect, i.e. it will try to decrease the concentration of the acid, $[H^+]$. It can only achieve this if the additional H^+ reacts with some of the CrO_4^{2-} to form the products $Cr_2O_7^{2-}$ and H_2O. Put in very simple terms, by adding acid we add to the left-hand side. The system counters this by moving to the right-hand side, and the colour changes to orange.

Effect of changing pressure on the position of the equilibrium

Pressure has virtually no effect on the reactions of solids or liquids and only affects gaseous reactions. The pressure of the gas mixture simply depends on the number of gas molecules in the mixture. The greater the number of gas molecules in the equilibrium mixture, the greater the pressure in the equilibrium mixture. If the pressure is increased, then the system under equilibrium will try to decrease the pressure by reducing the number of gas molecules in the system.

If the pressure is increased in a system such as $2SO_2(g) + O_2(g) \rightleftharpoons 2SO_3(g)$, the position of the equilibrium will move to the *right*, reducing the number of molecules. This has the effect of reducing the pressure.

If the pressure is increased in a system such as $N_2O_4(g) \rightleftharpoons 2NO_2(g)$, the position of the equilibrium will move to the *left*, reducing the number of molecules. This has the effect of reducing the pressure.

If the pressure is increased on a system such as $2HI(g) \rightleftharpoons H_2(g) + I_2(g)$, the position of the equilibrium will *not* move to the right or to the left, because there are the same number of molecules on each side of the equilibrium, and movement from one side to the other has no effect on the pressure.

Increasing pressure also increases the rate of reaction.

Effect of changing temperature on the position of the equilibrium

Temperature not only influences the rate of the reaction, but also plays an important role in determining the position of the equilibrium. The effect of temperature can only be predicted if the ΔH value of the reaction is known.

For the reaction:

$$2A(g) + B(g) \rightleftharpoons C(g) + D(g) \qquad \Delta H = -100\,kJ\,mol^{-1}$$

it follows that the forward reaction:

$$2A(g) + B(g) \rightarrow C(g) + D(g) \qquad \Delta H = -100\,kJ\,mol^{-1}$$

is exothermic, and the reverse reaction:

$$C(g) + D(g) \rightarrow 2A(g) + B(g) \qquad \Delta H = +100\,kJ\,mol^{-1}$$

is endothermic.

If we increase the temperature for the reaction mixture, then, according to le Chatelier's principle, the system will try to decrease the temperature. The equilibrium mixture can achieve this by favouring the reverse reaction, which is endothermic, and therefore removes the additional enthalpy caused by increasing the temperature. The position of the equilibrium moves to the left.

$2HI(g) \rightleftharpoons H_2(g) + I_2(g)$ is an *endothermic* reaction, and therefore if:

- the temperature is increased then the equilibrium moves to the right
- the temperature is decreased then the equilibrium moves to the left

$SO_2(g) + O_2(g) \rightleftharpoons 2SO_3(g)$ is an *exothermic* reaction, and therefore if:

- the temperature is increased then the equilibrium moves to the left
- the temperature is decreased then the equilibrium moves to the right

Effect of using a catalyst on the position of the equilibrium

The definition of a catalyst is a substance that speeds up the rate of reaction, without itself being changed, by lowering the activation energy of the reaction. In addition, the catalyst does not alter the amount of product produced.

In a system under equilibrium, a catalyst will speed up the forward and the reverse reactions equally, and will therefore have no effect on the position of the equilibrium. However, catalysts still play an important part in reversible reactions, as they reduce the time taken to reach equilibrium. In brief, a catalyst will produce the same amount of product, but will produce it more quickly.

The Haber process

We need large amounts of nitrogen compounds, particularly for fertiliser, and the Haber process is essential for the 'fixation' of atmospheric nitrogen. Atmospheric nitrogen is in plentiful supply, but cannot be used directly and must be converted (i.e. fixed) into a useful compound. The Haber process converts nitrogen into ammonia.

$$N_2(g) + 3H_2(g) \rightleftharpoons 2NH_3(g) \qquad \Delta H = -93\,kJ\,mol^{-1}$$

le Chatelier's principle allows us to predict the optimum conditions for this industrial process.

ΔH is $-93\,kJ\,mol^{-1}$, which tells us that the forward reaction is exothermic and, therefore, if:

- the temperature is increased then the equilibrium moves to the left
- the temperature is decreased then the equilibrium moves to the right

The optimum temperature to achieve the maximum yield of ammonia is therefore a low temperature.

If the pressure is increased on a system such as $N_2(g) + 3H_2(g) \rightleftharpoons 2NH_3(g)$, then the position of the equilibrium will move to the *right*, so that the number of molecules is reduced, which has the effect of reducing the pressure.

The optimum pressure to achieve the maximum yield of ammonia is therefore a high pressure.

Table 11 illustrates the effect of changing temperature and pressure on the yield of ammonia.

Temperature/K	Pressure/atm			
	25	50	100	200
373	92	94	96	98
573	28	40	53	67
773	3	6	11	18
973	1	2	4	9

Table 11 Percentage yield of ammonia

Low temperature (373 K = 100°C) gives the highest percentage yield. However, at low temperature the rate of reaction is slow, and a compromise has to be reached between yield and rate of reaction.

High pressure (200 atm) gives the highest percentage yield. High pressure also increases the rate of reaction. However, at high pressure the operating costs increase and a compromise has to be reached between yield, rate and cost.

Manufacturing conditions

The manufacture of ammonia in a modern plant is highly efficient. The operating conditions are a compromise:

- a temperature of around 700 K (427°C)
- a pressure of around 100 atm

The rate of reaction is greatly improved by using a finely divided or porous iron catalyst, which incorporates metal oxide promoters.

Knowledge check 15

Methanol, $CH_3OH(l)$, is manufactured by reacting carbon monoxide with hydrogen.

$CO(g) + 2H_2(g) \rightarrow CH_3OH(g)$ $\Delta H = -91\,kJ\,mol^{-1}$

a State the effect on the position of the equilibrium of:

i increasing the temperature

ii increasing the pressure

iii increasing the concentration of carbon monoxide

iv using a catalyst

b The catalyst used is a mixture of copper and zinc oxides. State whether the catalyst is homogeneous or heterogeneous.

Equilibrium constant, K_c

When a reversible reaction reaches equilibrium the concentrations of the reactants and products remain unchanged. By calculating the ratio of the concentrations of the products and the reagents we obtain an important constant, K_c, known as the equilibrium constant. The equilibrium constant varies with temperature but its value is the same for any other change in condition.

In general terms for the reaction:

$$aA + bB \rightleftharpoons cC + dD$$

the equilibrium constant, K_c, is given by:

$$K_c = \frac{[C]^c[D]^d}{[A]^a[B]^b}$$

In the expression for K_c the square brackets '[ ]' indicate that the concentrations of the reactants and products are expressed in units of $mol\,dm^{-3}$.

The products appear on the top line of the expression and the reactants on the bottom line. Each concentration term is raised to the power of the number in front of its formula in the balanced equation.

For the equilibrium:

$$3H_2(g) + N_2(g) \rightleftharpoons 2NH_3(g)$$

$$K_c = \frac{[NH_3(g)]^2}{[H_2(g)]^3\,[N_2(g)]}$$

For the equilibrium:

$$PCl_3(g) + Cl_2 \rightleftharpoons PCl_5(g)$$

$$K_c = \frac{[PCl_5(g)]}{[PCl_3(g)]\,[Cl_2(g)]}$$

The value of K_c gives an indication of the position in the equilibrium mixture. If K_c has a large value, it means that there are more of the products present than the reactants. If K_c is small, it is the reactants that are present in larger quantity. The values of K_c vary enormously from large values to values that are very small.

For example, in the equilibrium:

$$H_2(g) + CO_2(g) \rightleftharpoons CO(g) + H_2O(g)$$

$$K_c = \frac{[CO(g)]\,[H_2O(g)]}{[H_2(g)]\,[CO_2(g)]} = 7.8 \times 10^{-3} \text{ (no units) at } 500\,K$$

The value of K_c is small, which indicates that the equilibrium lies to the left and the concentration of the reactants is far bigger than that of the products.

For the equilibrium between sulfur dioxide, oxygen and sulfur trioxide:

$$2SO_2(g) + O_2(g) \rightleftharpoons 2SO_3(g)$$

the equilibrium constant:

$$K_c = \frac{[SO_3(g)^2]}{[SO_2(g)^2]\,[O_2(g)]} = 4.0 \times 10^{24}\,dm^3\,mol^{-1} \text{ at } 289\,K$$

The equilibrium constant is large, indicating that there is vastly more product (SO_3) than there are reactants.

Exam tip

You are not expected to supply units for equilibrium constants in the AS exam.

Effect on K_c of a change in temperature

The effect of a change in temperature on the value of K_c is best explained using examples of exo- and endothermic reactions (Table 12).

Exothermic	Endothermic
$2SO_2(g) + O_2(g) \rightleftharpoons 2SO_3(g)$ $\Delta H = -97\,kJ\,mol^{-1}$	$H_2(g) + CO_2(g) \rightleftharpoons CO(g) + H_2O(g)$ $\Delta H = +42\,kJ\,mol^{-1}$
If the temperature is increased the reverse endothermic reaction is favoured and the equilibrium moves to the **left**	If the temperature is increased the forward endothermic reaction is favoured and the equilibrium moves to the **right**
It follows that the value of the equilibrium constant will **decrease** as the temperature rises	It follows that the value of the equilibrium constant will **increase** as the temperature rises

Table 12

Calculating an equilibrium constant

If the equilibrium concentration of each component of an equilibrium mixture is known, the equilibrium constant can be calculated by substituting these values into the expression for the equilibrium constant.

Example

At 450°C, hydrogen and gaseous iodine form an equilibrium mixture with hydrogen iodide:

$$H_2(g) + I_2(g) \rightleftharpoons 2HI(g)$$

When an equilibrium mixture is analysed, it is found to contain 0.015 mol dm^{-3} of HI, 0.0012 mol dm^{-3} of I_2 and 0.0038 mol dm^{-3} of H_2. Calculate the value of the equilibrium constant to 2 significant figures.

$$K_c = \frac{[HI(g)]^2}{[H_2(g)]\,[I_2(g)]} = \frac{(0.015)^2}{(0.0012)\,(0.0038)} = 49$$

Knowledge check 16

Write the equilibrium constant for the equilibrium:

$2NO_2(g) \rightleftharpoons N_2O_4(g)$

a At a certain temperature the equilibrium constant for this reaction has a value of 0.0025. What does this indicate about the position of the equilibrium?

b If the concentration of NO_2 is $1.0\,mol\,dm^{-3}$, calculate the concentration of the N_2O_4 at this temperature ($K_c = 0.0025\,dm^3\,mol^{-1}$).

Knowledge check 17

For the reaction:

$H_2O(g) + C(s) \rightleftharpoons H_2(g) + CO(g)$

a Write an expression for the equilibrium constant K_c. (Hint: solids do not appear in the equation for K_c.)

b At 25°C the value of $K_c = 1.0 \times 10^{-16}\,mol\,dm^{-3}$.
At 250°C the value of $K_c = 2.4 \times 10^{-7}\,mol\,dm^{-3}$.

i At 25°C, explain whether or not the reaction lies to the left or to the right.

ii Explain whether or not the forward reaction is exothermic or endothermic.

Summary

Having revised **Module 3 Physical chemistry** you should now have an understanding of:

- exothermic and endothermic reactions
- enthalpies of reaction, formation and combustion
- bond enthalpies
- Hess's law and calculations using enthalpy cycles
- Boltzmann distributions and how temperature and catalysts affect rate
- catalysts
- le Chatelier's principle
- the equilibrium constant K_c

Module 4 Core organic chemistry

Basic concepts and hydrocarbons

Basic concepts of organic chemistry

Nomenclature should follow IUPAC rules for naming organic compounds. The IUPAC (International Union of Pure and Applied Chemistry) rules for naming are based around the systematic names given to the alkanes and the prefix or suffix given to each functional group.

In this module you will be expected to recognise and to name alkanes, alkenes, alcohols and haloalkanes.

- Alkanes have a general formula of C_nH_{2n+2}.
- Alkenes have a general formula of C_nH_{2n}.
- Alcohols have a general formula of $C_nH_{2n+1}OH$.
- Haloalkanes have a general formula of $C_nH_{2n+1}X$, where X = F, Cl, Br or I.

It is important that you learn the appropriate prefix or suffix for each group:

- Alkanes always end in '-ane'.
- Alkenes always end in '-ene'.
- Alcohols always end in '-ol'.
- Haloalkanes always start with 'fluoro-', 'chloro-', 'bromo-' or 'iodo-'.

In Figure 18:

- **A** is an alkane and hence ends in *...ane*. The longest carbon chain is five, hence *pent*ane. There is a branch (on the second carbon atom) that consists of one carbon, hence *2-methyl*, giving the full name *2-methylpentane*.
- **B** is an alkene and hence ends in *...ene*. The longest carbon chain is four, hence *butene*. The double bond starts at carbon atom one, hence but-*1*-ene. There is a branch (on the second carbon atom) that consists of one carbon, hence *2-methyl*, giving the full name of *2-methylbut-1-ene*.
- **C** is both an alcohol (and hence ends in *...ol*) and a bromoalkane (and hence starts with *bromo...*). The longest carbon chain is six, hence *hex*ane. The alcohol group is on the second carbon (hence *-2-ol*) and the bromine is on the fourth carbon (hence *4-bromo...*). The full name is *4-bromohexan-2-ol*.

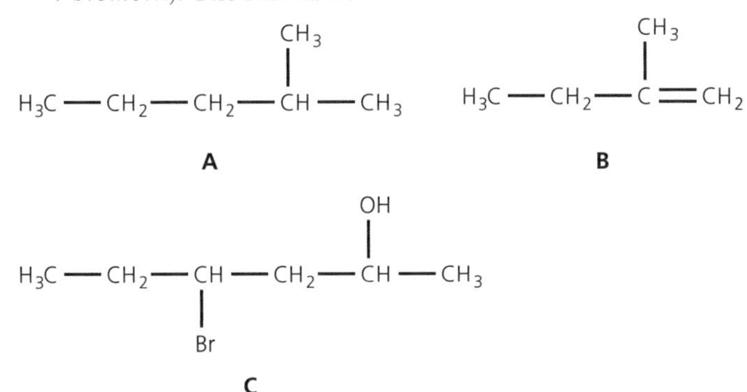

Figure 18

You will be expected to draw and to use structural formulae, displayed formulae and skeletal formulae.

A **structural formula** is accepted as the minimal detail, using conventional groups, for an unambiguous structure. The molecular formula C_4H_{10} could be either of two isomers, butane or methylpropane, and is therefore ambiguous. The structural formula for butane is $CH_3CH_2CH_2CH_3$ which could also be written as $CH_3(CH_2)_2CH_3$, and the structural formula of methylpropane is written as $(CH_3)_3CH$.

A **displayed formula** shows both the relative placing of atoms and the number of bonds between them. The displayed formulae for butane and methylpropane are shown in Figure 19.

Knowledge check 18

Name the following compounds:

a $CH_3CHBrCH_3$
b $CH_3CHCHCH_3$
c $(CH_3)_4C$
d $CH_3CH_2CH(CH_3)CH_2CH_3$

Knowledge check 19

Draw the displayed formula for each of 2-chloropropane, butan-2-ol and 3-methylbut-1-ene.

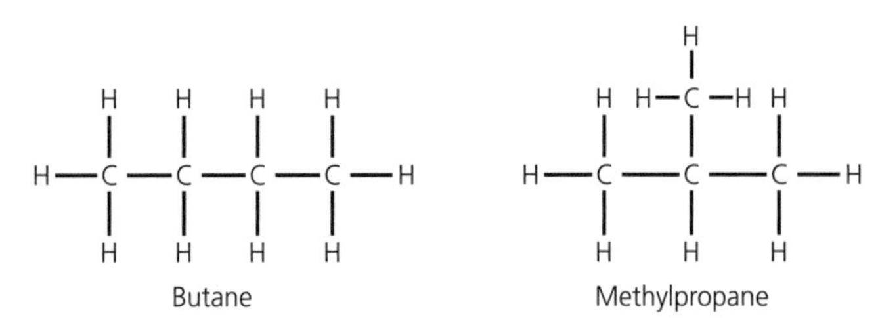

Figure 19

A **skeletal formula** is used to show a simplified organic formula by removing hydrogen atoms from alkyl chains, leaving only the carbon skeleton and associated functional groups. The skeletal formulae for butane and methylpropane are shown in Figure 20 with other examples.

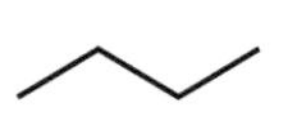
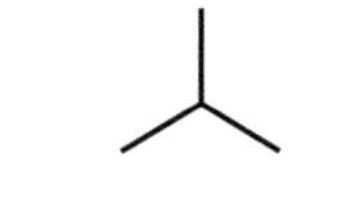
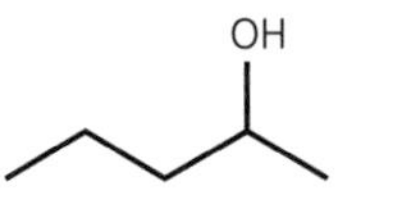

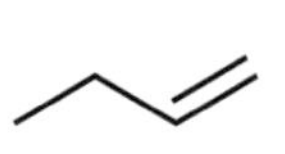

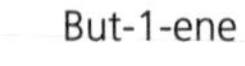

Figure 20

Cyclic compounds such as cyclohexane and benzene are represented as shown in Figure 21.

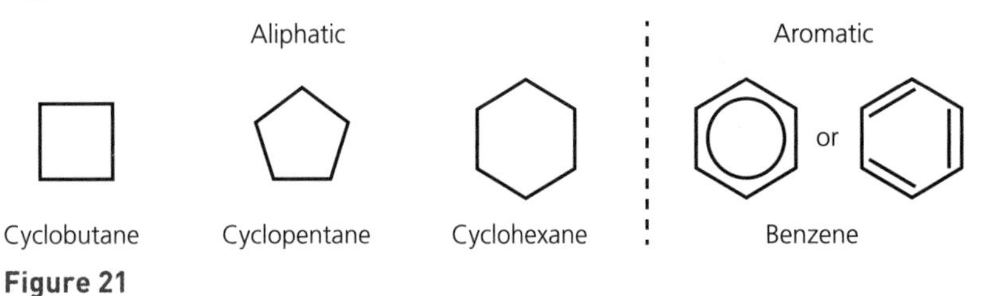

Figure 21

Knowledge check 20

Draw the skeletal formula for each of pent-2-ene, butan-2-ol and 3-chlorobut-1-ene.

Isomerism

Structural isomerism

Structural isomers are defined as compounds with the same molecular formula but different structural formulae. They occur with all functional groups, and you may be expected to both draw and name them. Figure 22 shows an example.

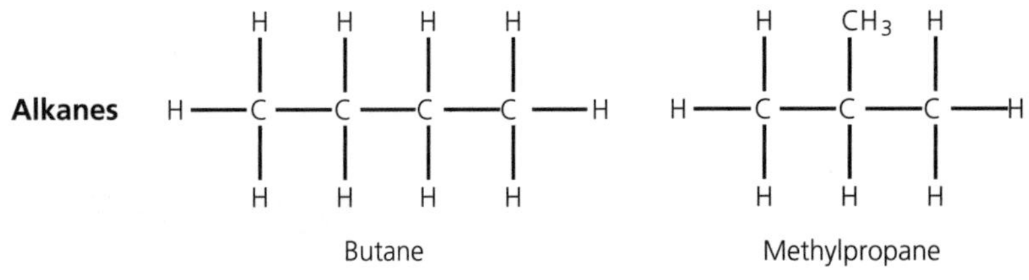

Figure 22

E/Z (or cis–trans) isomerism

E/Z isomerism is found in alkenes. The key features to look for are:

- the C=C double bond
- each C in the C=C double bond is bonded to two different atoms or groups

The C=C double bond ensures that there is restricted rotation about a double bond, and the different atoms or groups attached to each carbon atom ensure that there is no symmetry around each individual carbon atom in the C=C double bond.

Exam tip

When explaining what is meant by *E/Z* isomerism most students remember that the C=C double bond restricts rotation but lots fail to mention the requirement that each C in the C=C double bond must be attached to two different atoms or groups.

Calculations

Most chemistry examinations will contain some calculations. The sort of calculations that could be tested within the organic chemistry section are: empirical and molecular formulae, percentage yield and atom economy calculations all of which were covered in the student guide covering Modules 1 and 2 of this series.

Knowledge check 21

A student reacted 5.00 g of methanol (CH_3OH) with 9.38 g of ethanoic acid (CH_3COOH) and made 5.20 g of methyl ethanoate (CH_3COOCH_3). The equation for the reaction is:

$$CH_3COOH + HOCH_3 \rightleftharpoons CH_3COOCH_3 + H_2O$$

5.00 g   5.20 g

Calculate:

a the student's percentage yield
b the atom economy

Exam tip

When carrying out calculations do not round numbers during the calculation. Keep the numbers in your calculator and only round when you have finished the entire calculation.

Reactions of functional groups

When describing the reactions of any functional group, you will be expected to:

- know the reagents
- know the conditions
- be able to write balanced equations
- be able to give the mechanism

Definitions

Reagents: these are chemicals involved in the reaction.

Conditions: these normally comprise the temperature, pressure and/or the use of a catalyst.

Mechanism: this breaks down the overall reaction into separate steps. It is usual to identify the attacking species. This is a radical, an electrophile or a nucleophile.

Radicals: reactive particles with an *unpaired* electron. The symbol for a radical generally shows the unpaired electron as a dot, for example Cl•, CH_3•.

Electrophiles: these are electron-deficient and can accept a lone pair of electrons — for example H^+, NO_2^+.

Nucleophiles: these are molecules or ions that can donate a lone pair of electrons — for example OH^-, NH_3.

Aliphatic: these are compounds that contain hydrocarbon chains, branched chains or non-aromatic rings.

Aromatic: compounds that contain a benzene ring.

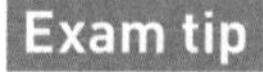

Exam tip

Words that end with '... phile' indicate a liking for. An 'anglophile' is someone who likes England. Don't be tempted to do this when explaining key terms like electrophile. You will *not* score any marks for writing that it 'loves electrons'. Stick to the scientific definitions given above.

Hydrocarbons: alkanes

Physical properties of alkanes

Hydrocarbons are compounds that contain hydrogen and carbon *only*. Alkanes and cycloalkanes are saturated hydrocarbons, as all the C–C bonds are single bonds. These bonds result in a tetrahedral shape, with bond angles of 109.5° around each carbon atom. Alkanes are obtained from crude oil, which is a complex mixture of hundreds of different hydrocarbons. The mixture undergoes initial separation by fractional distillation. Separation is achieved because the different alkanes have different boiling points.

The variation in boiling points depends on the amount of intermolecular bonding. Intermolecular bonding is covered in depth in the student guide covering Modules 1 and 2 of this series. Alkanes are molecules with very low or no polarity and hence the intermolecular forces present are induced dipole–dipole interactions (van der Waals forces).

There are two important trends in the variation of boiling points in alkanes. First, as the relative molecular mass increases, the boiling point increases due to:

- an increase in chain length
- an increase in the number of electrons

Both of the above result in an increase in the number of induced dipole–dipole forces.

Second, for isomers with the same relative molecular mass, the boiling points decrease with an increase in the amount of branching. This can be explained by the fact that straight chains pack closer together, creating more intermolecular forces. There are three isomers with the formula C_5H_{12} (Figure 23).

Pentane:
boiling point = 36°C

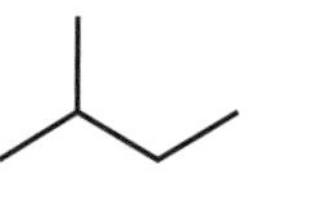

Methylbutane:
boiling point = 28°C

Dimethylpropane:
boiling point = 10°C

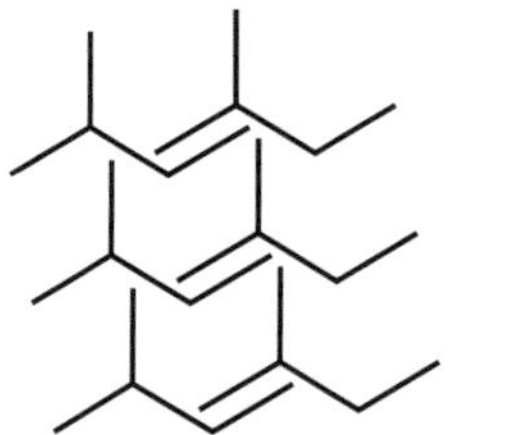

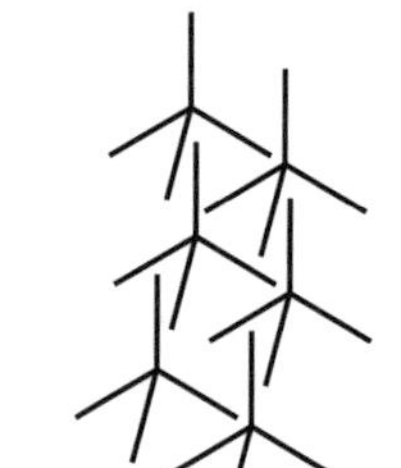

Figure 23

Knowledge check 22

Put the following alkanes in order of boiling points starting with lowest boiling point first: octane, 2-methylpentane, hexane and 2,3-dimethylbutane.

Reactions of alkanes

Alkanes are relatively unreactive, because:

- the C–C and the C–H bonds in alkanes are strong bonds
- the alkanes are either non-polar or have very low polarity

Combustion of alkanes

Alkanes burn easily and release energy (i.e. they undergo an exothermic reaction), and are used as fuels in industry, the home and in transport.

Complete combustion of alkanes in an excess of oxygen produces carbon dioxide and water.

$CH_4 + 2O_2 \rightarrow CO_2 + 2H_2O$

$C_2H_6 + 3½O_2 \rightarrow 2CO_2 + 3H_2O$

Incomplete combustion of alkanes in a limited supply of oxygen produces carbon monoxide and water.

$CH_4 + 1½O_2 \rightarrow CO + 2H_2O$

$C_2H_6 + 2½O_2 \rightarrow 2CO + 3H_2O$

Carbon monoxide is poisonous, and it is essential that hydrocarbon fuels are burnt in a plentiful supply of oxygen. Cars are fitted with catalytic converters (see p. 25) to ensure that the amount of carbon monoxide emitted is reduced.

Substitution reactions of alkanes

Substitution by bromine and by chlorine to form haloalkanes

Reaction between methane and bromine

Reagent: Br_2

Conditions: ultraviolet light

Equation: $CH_4 + Br_2 \rightarrow CH_3Br + HBr$

Mechanism: the overall reaction is a **radical substitution**

Initiation: $Br_2 \rightarrow 2Br\bullet$ This requires ultraviolet light and the bond breaks by homolytic fission (homolysis)

Propagation 1: $CH_4 + Br\bullet \rightarrow HBr + CH_3\bullet$

Propagation 2: $CH_3\bullet + Br_2 \rightarrow CH_3Br + Br\bullet$

Termination: any two free radicals

$CH_3\bullet + CH_3\bullet \rightarrow C_2H_6$ *or*

$CH_3\bullet + Br\bullet \rightarrow CH_3Br$ *or*

$Br\bullet + Br\bullet \rightarrow Br_2$

There are three distinct stages to the mechanism:

1 **Initiation:** radicals are generated. The ultraviolet light provides sufficient energy to break the Br–Br bond by **homolytic fission** and generates **radicals** (Figure 24).

Br—Br ⟶ Br• + •Br

Figure 24

Homolytic fission is the breaking of a covalent bond such that each atom gets one of the two shared electrons.

A **radical** is a particle that contains a single unpaired electron.

2 **Propagation:** this involves two steps, each one maintaining the radical concentration. Usually, a chlorine or bromine radical is swapped for an alkyl radical or vice versa.

3 **Termination:** this involves the loss of radicals.

Radical reactions have limitations, in that it is almost impossible to produce a single product. This is because radicals are very reactive and it is very difficult to:

- avoid multiple substitutions of the hydrogen atoms in the alkane. Typically, for the reaction between CH_4 and Cl_2, the products would consist of CH_3Cl, CH_2Cl_2, $CHCl_3$ and CCl_4.
- determine the position of substitution. In the mono-substitution of pentane using $Cl_2(g)$ would produce a mixture of the three isotopes 1-chloropentane, 2-chloropentane and 3-chloropentane.

This makes separation difficult and costly.

Exam tip

Remember that the first propagation step *always* produces a hydrocarbon radical ($•C_nH_{2n+1}$) and HBr (or HCl).

Knowledge check 23

2-methylpentane reacts, in the presence of ultraviolet light, with $Br_2(l)$ to produce a mixture of isomers all with molecular formula $C_6H_{13}Br$. List, by name, each isomer that could be formed.

Hydrocarbons: alkenes

Alkenes and cycloalkenes are unsaturated hydrocarbons and contain at least one C=C double bond. The double bond is made up of a σ-bond and a π-bond.

A **σ-bond** is a single covalent bond made up of two shared electrons with the electron density concentrated between the two nuclei.

A **π-bond** is formed by the sideways overlap of two adjacent *p*-orbitals (Figure 25).

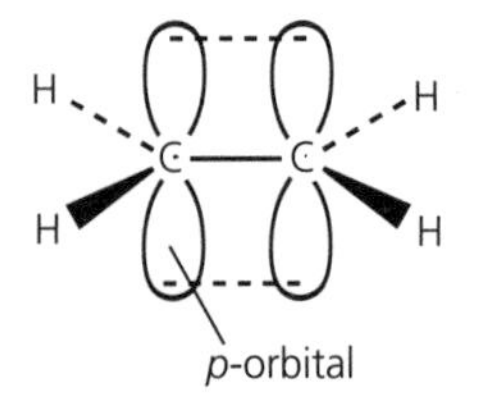

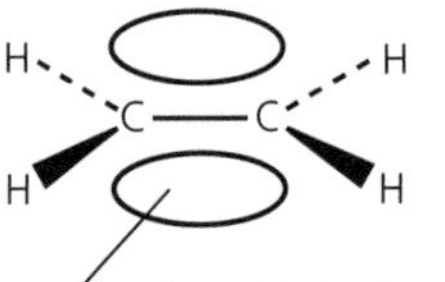

Figure 25

The bond angle at each side of the C=C double bond is approximately 120° (usually in the region 116–124°), which results in a trigonal planar structure around the C=C double bond (Figure 26).

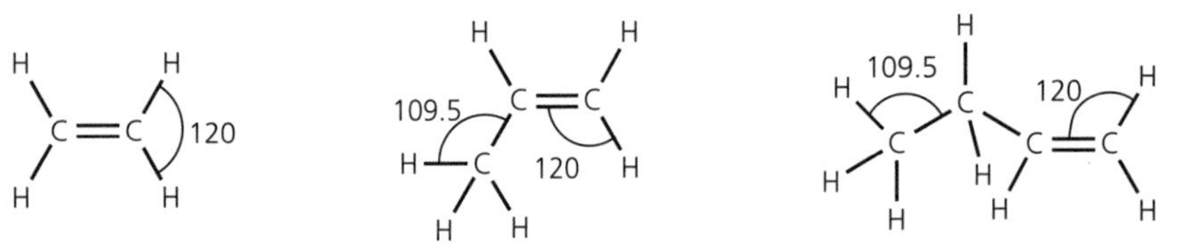

Figure 26

Exam tip

When explaining *E/Z* isomerism makes sure you use the right wording. 'Each carbon atom in the C=C double bond is bonded to two different atoms or *molecules*' will not score the mark.

The C=C double bond prevents freedom of rotation, which under certain circumstances can lead to the existence of stereoisomers known as *E*/*Z* (*cis–trans*) isomers. The key features essential for *E*/*Z* isomers are:

- the C=C double bond
- each carbon atom in the C=C double bond is bonded to two different atoms or groups

It is important that you can recognise or predict whether or not *E*/*Z* isomers exist.

Stereoisomerism in alkenes

There is a set of rules known as the Cahn, Ingold and Prelog (CIP) rules that enable you to decide whether you should name each isomer as an **E isomer** or a **Z isomer**.

First decide whether or not a compound can exist as *E*/*Z* isomers by asking yourself two simple questions:

1. Does it have a C=C double bond?
2. Is each C in the C=C double bond bonded to two different atoms or groups?

If the answer to both these questions is 'yes' then you need to follow the CIP rules to determine whether it is either the *E* isomer or the Z isomer.

Cahn, Ingold and Prelog (CIP) Rules

These rules are shown in Table 13.

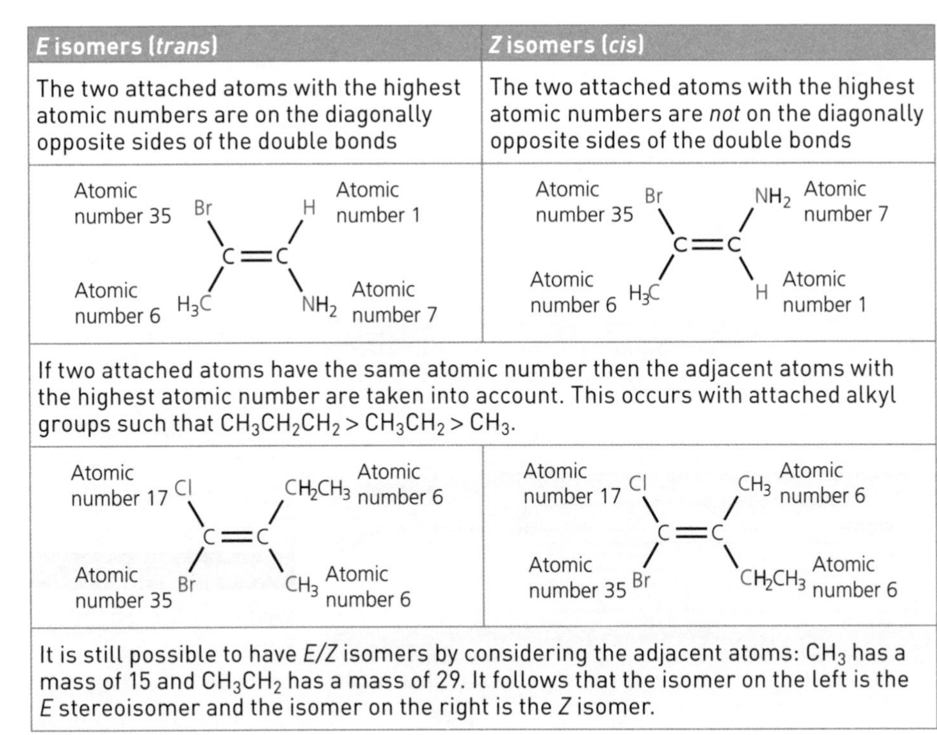

***E* isomers (*trans*)**	***Z* isomers (*cis*)**
The two attached atoms with the highest atomic numbers are on the diagonally opposite sides of the double bonds	The two attached atoms with the highest atomic numbers are *not* on the diagonally opposite sides of the double bonds
Atomic number 35 Br, H Atomic number 1; C=C; Atomic number 6 H_3C, NH_2 Atomic number 7	Atomic number 35 Br, NH_2 Atomic number 7; C=C; Atomic number 6 H_3C, H Atomic number 1
If two attached atoms have the same atomic number then the adjacent atoms with the highest atomic number are taken into account. This occurs with attached alkyl groups such that $CH_3CH_2CH_2 > CH_3CH_2 > CH_3$.	
Atomic number 17 Cl, CH_2CH_3 Atomic number 6; C=C; Atomic number 35 Br, CH_3 Atomic number 6	Atomic number 17 Cl, CH_3 Atomic number 6; C=C; Atomic number 35 Br, CH_2CH_3 Atomic number 6
It is still possible to have *E*/*Z* isomers by considering the adjacent atoms: CH_3 has a mass of 15 and CH_3CH_2 has a mass of 29. It follows that the isomer on the left is the *E* stereoisomer and the isomer on the right is the *Z* isomer.	

Table 13

Cis–trans isomerism is a special case of *E*/*Z* isomerism in which two of the substituent groups attached to each carbon atom of the C=C group are the same.

Exam tip

When asked to describe how the π-bond is formed in an alkene, students sometimes draw the following, which is incorrect — this shows a treble (not double) bond between the two carbons.

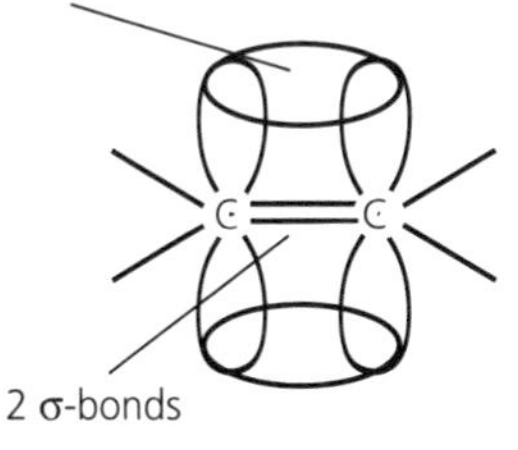

***E*/*Z* isomerism** *E*/*Z* isomers have the same structural formula but the atoms are arranged differently in space around a C=C double bond.

Knowledge check 24

Draw and name structural and stereoisomers of C_5H_{10}.

Addition reactions of alkenes

Alkenes are relatively reactive, because:

- the C=C bond consists of a σ-bond and a π-bond; the π-bond is weak and easily broken
- the C=C bond contains four electrons and has high electron density

The C=C double bond is an unsaturated bond and therefore undergoes addition reactions. Essentially, the double bond opens and an atom or group is added to each of the carbons. The general reaction can be summarised as Figure 27.

$CH_2{=}CH_2 + X{-}Y \rightarrow CH_2XCH_2Y$

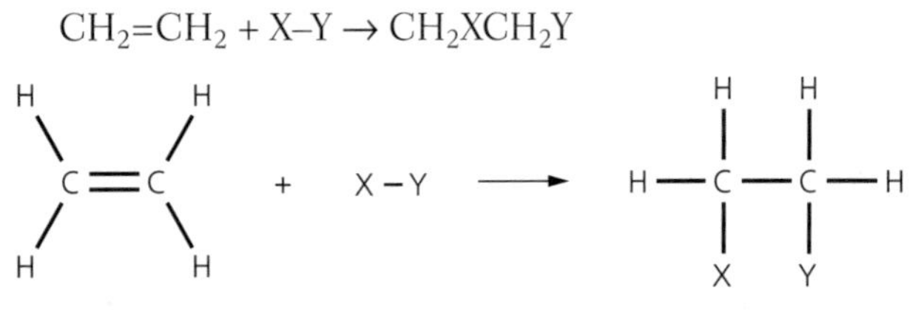

Figure 27

When preparing for the exam, it is good practice to stick to a routine. For most organic reactions, it is useful to know the reagents, conditions (if any), observations and the overall equation.

Hydrogenation

Reagents	Conditions (if any)	Observations (if any)	Overall equation
H_2	Ni catalyst, 150°C	None	$CH_2{=}CH_2 + H_2 \rightarrow CH_3CH_3$

Hydrogenation of polyunsaturated vegetable oils derived from plants is used in the production of some margarines. Similar conditions are used, so that vegetable oils react with hydrogen in the presence of a nickel catalyst.

Bromination

Reagents	Conditions (if any)	Observations (if any)	Overall equation
Br_2	None	Decolorisation of Br_2	$CH_2{=}CH_2 + Br_2 \rightarrow CH_2BrCH_2Br$

Formation of haloalkanes

Reagents	Conditions (if any)	Observations (if any)	Overall equation
HBr	None	None	$CH_2{=}CH_2 + HBr \rightarrow CH_3CH_2Br$

Hydration

Reagents	Conditions (if any)	Observations (if any)	Overall equation
H_2O (steam)	H_3PO_4 catalyst, 300°C, 6 MPa	None	$CH_2{=}CH_2 + H_2O \rightarrow CH_3CH_2OH$

All other alkenes undergo similar reactions under similar conditions. However, unsymmetrical alkenes, such as propene, can produce two isomers when reacted with hydrogen chloride (Figure 28) or water (Figure 29).

Exam tip

When asked to describe what you would see when bromine reacts with an alkene, many students lose the mark by stating that the bromine would go 'clear'. Clear is the wrong word. Bromine is already clear — it is a clear red/brown liquid. When it reacts it loses its colour and the correct word to use is 'decolorised'.

Knowledge check 25

Write a balanced equation for the reaction between buta-1,3-diene and bromine, state what you would observe and name the organic product.

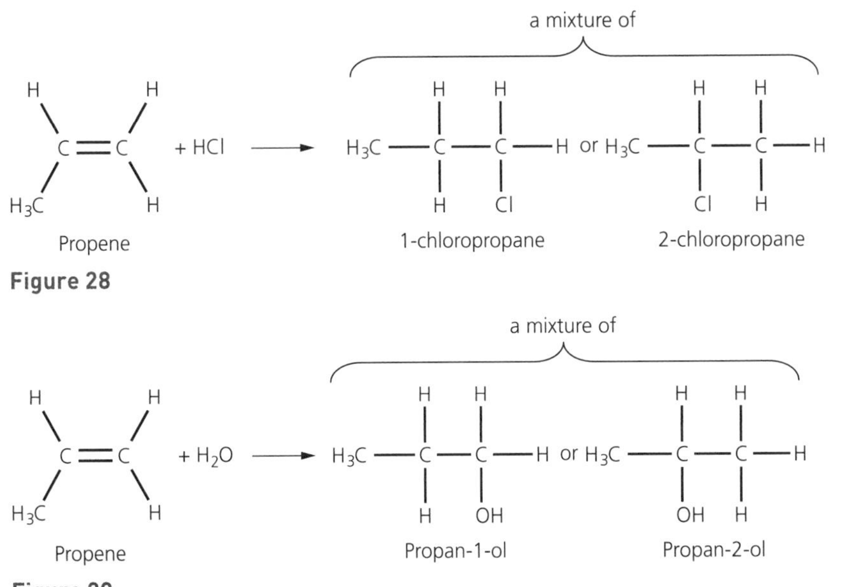

Figure 28

Figure 29

Electrophilic addition

Mechanisms involving electrophiles and nucleophiles involve the movement of electron pairs. This movement is shown by the use of curly arrows. The curly arrow always points from areas that are electron-rich to areas that are electron-deficient.

When describing mechanisms, it is essential that you show:

- relevant dipoles
- lone pairs
- curly arrows

Alkenes, such as ethene, undergo electrophilic addition reactions. An electrophile is defined as a lone pair (of electrons) acceptor.

The key features to the mechanism are as follows:

- When the Br–Br approaches the ethene, a temporary induced dipole is formed, resulting in $Br^{\delta+}$–$Br^{\delta-}$.
- The initial curly arrow starts at the π-bond (within the C=C double bond) and points to the $Br^{\delta+}$.
- The second curly arrow shows the movement of the bonded pair of electrons in the Br–Br to the $Br^{\delta-}$, resulting in heterolytic fission of the Br–Br bond.
- The formation of an intermediate carbonium ion (also called a carbocation) and a $:Br^-$ ion (that now contains the pair of electrons that were in the Br–Br bond) occurs.
- The third curly arrow from the $:Br^-$ to the positively charged carbonium ion results in the formation of 1,2-dibromoethane.

The mechanism is best described using curly arrows, dipoles and relevant lone pairs of electrons (Figure 30).

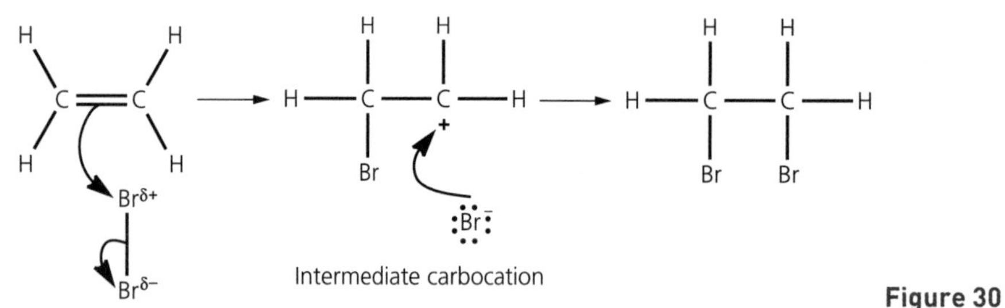

Figure 30

The intermediate ion can also be drawn as

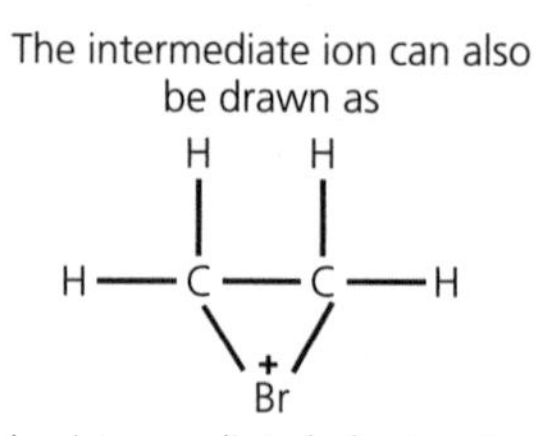

(an intermediate halonium ion)

When Br_2 reacts with an alkene, the Br–Br undergoes heterolytic fission (Figure 31). Compare this to the way the Br–Br bond is broken when it reacts with an alkane (see p. 37).

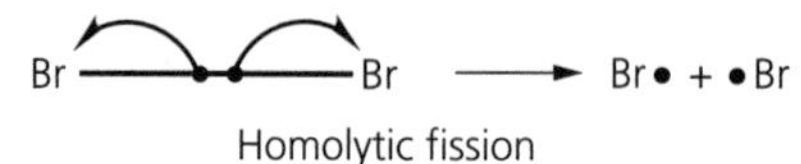

Figure 31

Exam tip

Mechanisms are generally well understood but it is easy to lose lots of marks by rushing and/or being careless. Look at the response below for the mechanism of the reaction between propene and Br_2. At first glance it looks good *but* there are *seven* errors or omissions. Can you spot all seven?

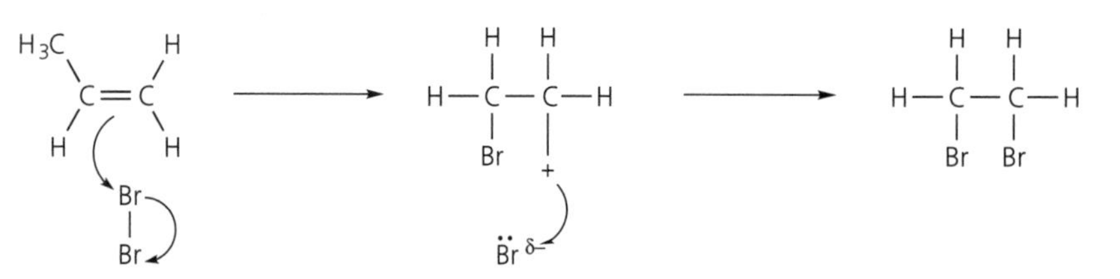

If you spot the errors made by others, it should prevent you from making the same mistakes.

Knowledge check 26

a Explain what is meant by an *electrophile*. Write a balanced equation for the reaction between Br_2 and cyclohexene.

b Describe, with the aid of curly arrows, the mechanism of the reaction between but-2-ene and HBr(g). Show relevant dipoles and lone pairs of electrons.

Markownikoff's rule

When an unsymmetrical reagent, usually either H–Br or H_2O but it could be H–Cl or H–CN, reacts with an unsymmetrical alkene such as propene, a mixture of two organic products, 1-bromopropane and 2-bromopropane, is formed (Figure 32).

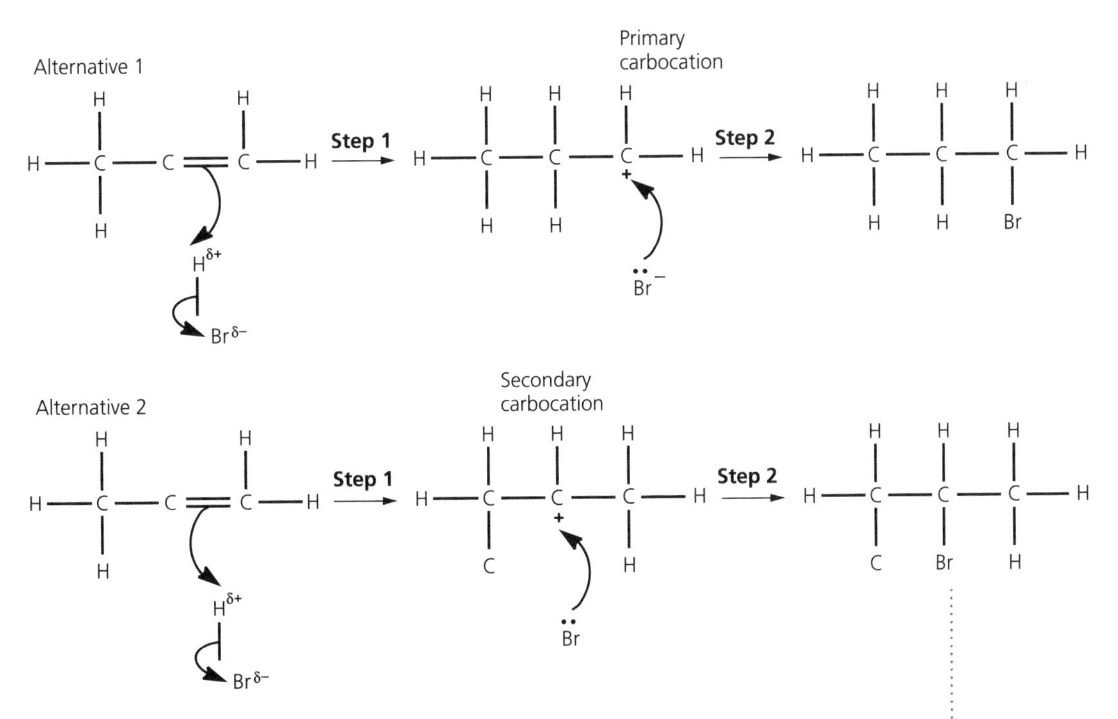

Figure 32

The major product, 2-bromopropane, can be predicted using **Markownikoff's rule**.

When propene, $CH_3CH{=}CH_2$, reacts with HBr — the *'addent other than hydrogen'* is Br and this will bond to the C, in the C=C double bond, with the least number of Hs (Figure 33). The Br will therefore bond to the CH (only has 1H) and not to the CH_2.

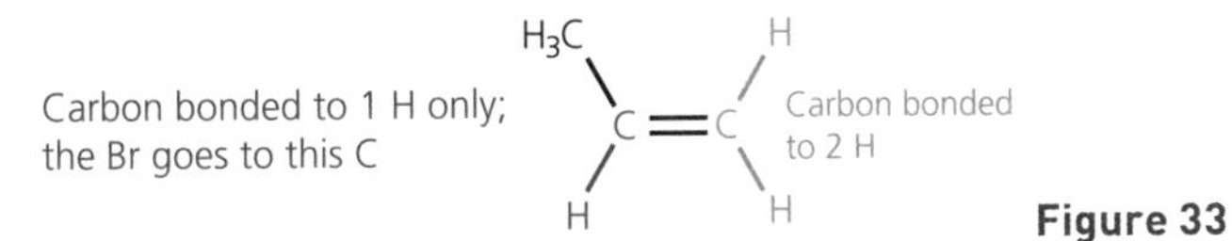

Figure 33

The product depends on the relative stabilities of the carbocation intermediates in the mechanism:

most stable — tertiary carbocation > secondary carbocation > primary carbocation — **least stable**

The more stable the carbocation intermediate, the more likely it will result in the product such that 'alternative 2' above is the favoured mechanism and results in the formation of 2-bromopropane.

Polymers from alkenes

Addition polymerisation

Alkenes can undergo an addition reaction in which one alkene molecule joins to another until a long molecular chain is built up. The individual alkene molecule is referred to as a **monomer**, while the long-chain molecule is known as the **polymer**.

Markownikoff's rule states that the addent other than hydrogen goes to the least hydrogenated carbon.

Knowledge check 27

Predict the major product formed when methylpropene reacts with steam in the presence of a suitable acid catalyst. Explain your answer.

Some common monomers and their reactions are shown in Figure 34.

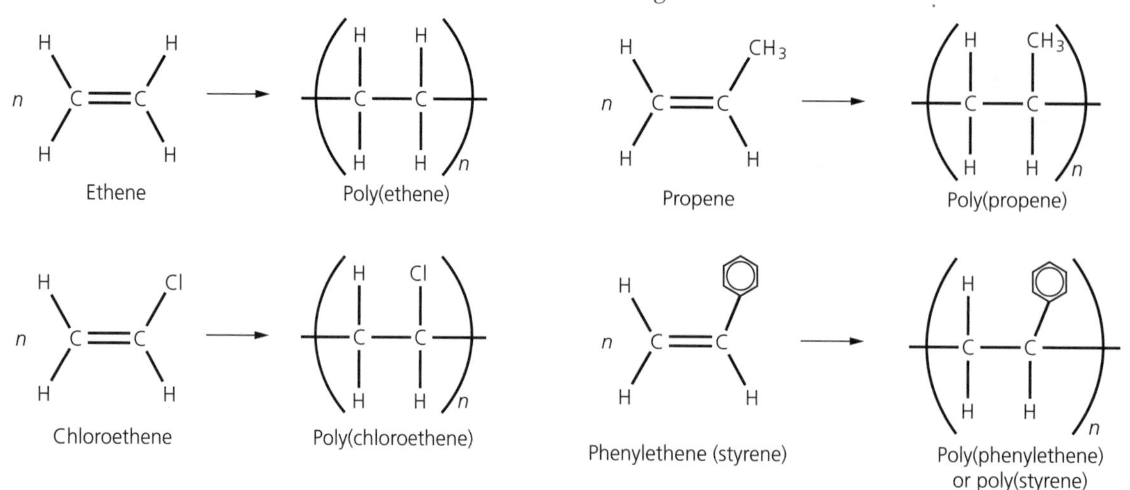

Figure 34

n is a large number and can be as big as 10 000. It is possible to deduce the repeat unit of an addition polymer and to identify the monomer from which the polymer was produced (Figure 35).

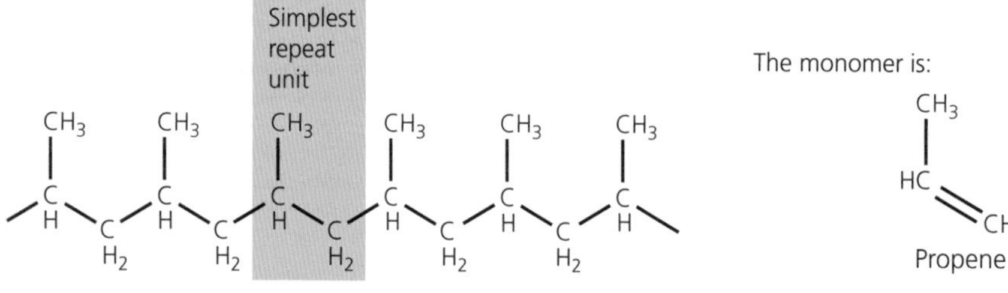

Figure 35

Exam tip

If asked to draw two repeat units of the polymer formed from propene, a common incorrect response is:

It is worth remembering that two repeat units of any addition polymer will always have a central backbone containing *four* carbon atoms.

Knowledge check 28

Write an equation to show the polymerisation of but-1-ene. Draw two repeat units of the polymer.

Waste polymers and alternatives

The widespread use of polymers made from alkenes creates two major problems.

1 The depletion of finite resources — polymers from alkenes use fractions obtained from crude oil.

2 Disposal — Plastic waste has for many years been buried in landfill sites, where it remains unchanged for decades. This means that local authorities have to find more and more landfill sites. The bonds in addition polymers are strong covalent bonds and are non-polar, making most of the polymers resistant to chemical attack. They are not broken down by bacteria and are referred to as being **non-biodegradable**.

An alternative to dumping is incineration. Polymers are hydrocarbon-based and are therefore potentially good fuels. When burnt, they release useful energy. In the UK only about 10% of plastic is incinerated, but in other countries, such as Japan and Denmark, as much as 70% is burnt to create useful energy. Some plastics, such as PVC, also produce toxic gases (e.g. HCl), so the incinerators have to be fitted with gas-scrubbers.

Another possibility is recycling the polymers and using them as feedstock for the production of new polymers. Different types of polymer have to be separated from each other, as a mixture of polymers, when recycled, produces an inferior plastic product.

Biodegradable and photodegradable polymers are another possibility, and considerable effort has gone into trying to develop those that have suitable properties. The general principle is to create a polymer containing an active functional group that can be attacked by bacteria. Promising options are based on the polymerisation of isoprene, which is the monomer from which rubber is made. The systematic name for isoprene is 2-methylbuta-1,3-diene (Figure 36).

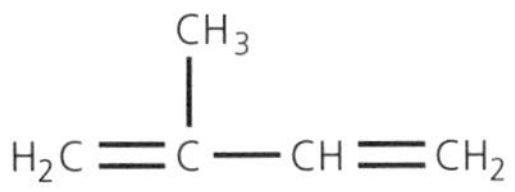

$$H_2C{=}C(CH_3){-}CH{=}CH_2$$

Figure 36

Other options are based on condensation polymers, which you will meet if you study chemistry further in the second year of the course. A polymer such as poly(lactic acid) is both biodegradable and renewable. It is biodegradable because it is readily broken down by bacteria. It is renewable because the monomer is obtained from plants such as corn or sugarcane.

Summary

Having revised **Module 4: Basic concepts and hydrocarbons** you should now have an understanding of:

- the various ways in which organic formulae can be represented
- isomerism
- calculations used in organic chemistry
- key terms used in organic chemistry
- bonding, shape and boiling points of alkanes
- hydrocarbons as fuels including fractional distillation, cracking, isomerisation and reforming
- reactions of alkanes
- radical substitution mechanism
- bonding and shape of alkenes
- isomerism including *E*/*Z* isomers, CIP rules
- addition reactions
- electrophilic addition mechanism, Markownikoff's rule
- polymerisation and the environmental aspects of waste polymers

Alcohols, haloalkanes and analysis

Alcohols

Properties

Alcohols all contain the hydroxyl group, –OH, and their names all end in '-ol'. The $O^{\delta-}$–$H^{\delta+}$ bond is polar and results in the formation of intermolecular hydrogen bonds.

Alcohols have relatively high boiling points (low volatility) and are miscible with water. This can be explained in terms of hydrogen bonding. Hydrogen bonds can be formed between the oxygen in the OH group in one alcohol and the hydrogen in the OH group in an adjacent alcohol (Figure 37). Hydrogen bonding decreases the volatility, resulting in an increase in boiling point. Methanol and ethanol are freely miscible with water. When mixed, some of the hydrogen bonds in the separate liquids are broken, but they are then replaced by new hydrogen bonds between the alcohol and water. The higher the relative molecular mass of the alcohol, the lower is its miscibility with water.

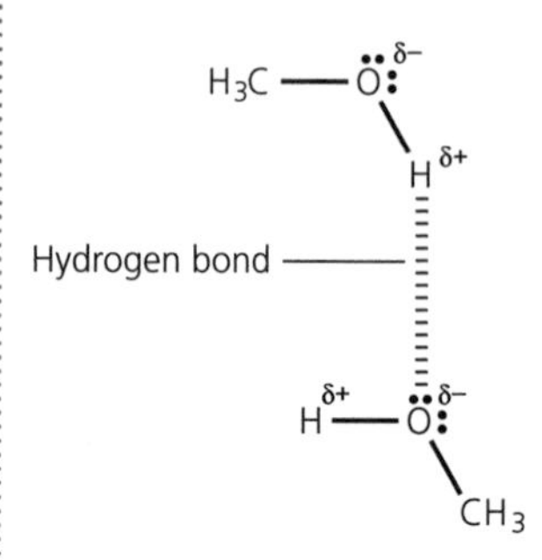

Figure 37

Alcohols can be classified as primary, secondary or tertiary. Using R to represent any other attachment, we can identify the nature of the alcohol (Figure 38).

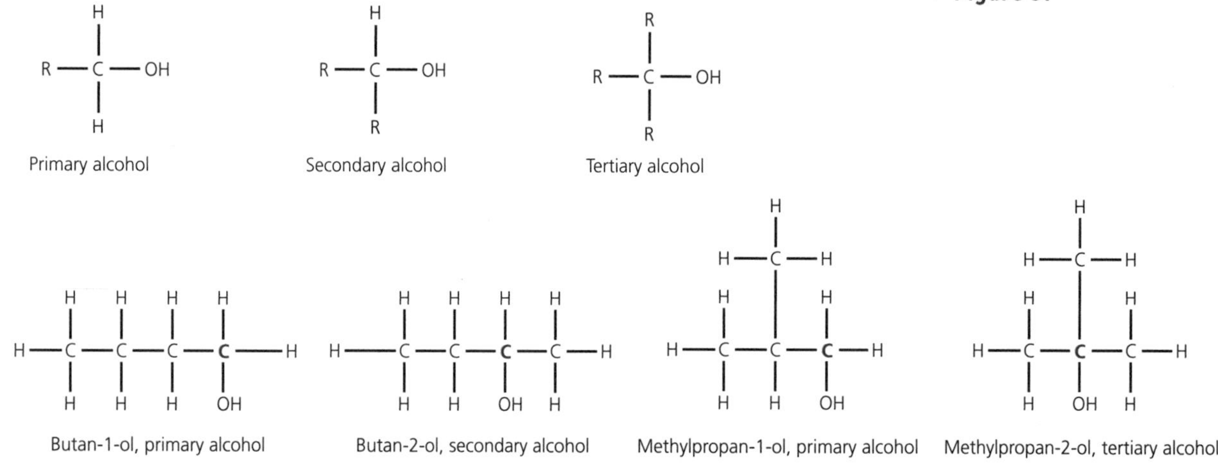

Figure 38

Reactions of alcohols

Exam tip

When drawing alcohols you have to be careful how you draw the bond to the hydroxyl (OH) group. It is easy to lose the mark as shown in A and in B below. C shows how it should be drawn.

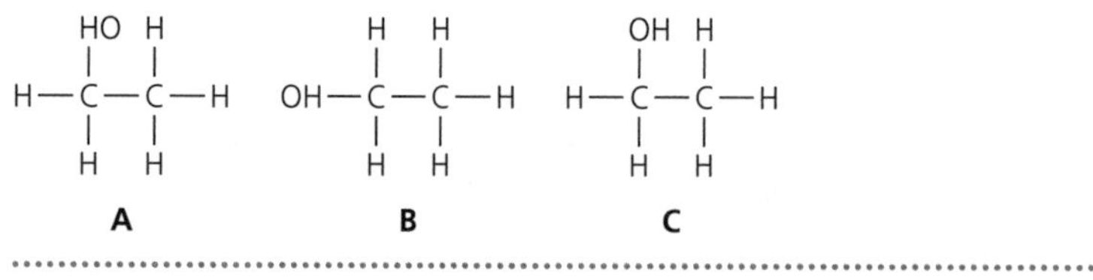

Combustion to produce carbon dioxide and water

Reagents	Conditions (if any)	Observations (if any)	Overall equation
Excess O_2	None	N/A	$C_2H_5OH + 3O_2 \rightarrow 2CO_2 + 3H_2O$

Dehydration to form an alkene

Reagents	Conditions (if any)	Observations (if any)	Overall equation
Concentrated H_2SO_4	High temperature (approx. 170°C)	None	$C_2H_5OH \rightarrow C_2H_4 + H_2O$

For alcohols like butan-2-ol it is possible to lose water in two ways (Figure 39).

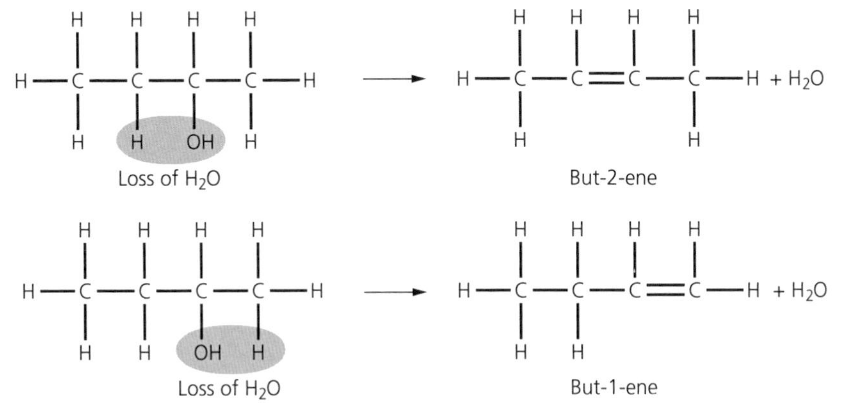

Figure 39

Oxidation

Oxidation reactions differ, depending on whether the alcohol is primary, secondary or tertiary.

- Primary alcohol → aldehyde → carboxylic acid
- Secondary alcohol → ketone
- Tertiary alcohol → resistant to oxidation
- Reagents: oxidising mixture $Cr_2O_7^{2-}/H^+$ (e.g. $K_2Cr_2O_7/H_2SO_4$)
- Conditions: when oxidising a primary alcohol, the choice of apparatus is important — refluxing will produce a carboxylic acid, while distillation produces an aldehyde
- Observations: each oxidation reaction is accompanied by a distinctive colour change from orange to green
- Balanced equations ([O] is used to represent the oxidising agent), see below.

Oxidation of primary alcohols to aldehydes

$CH_3OH + [O] \rightarrow HCHO + H_2O$

methanol methanal

$CH_3CH_2OH + [O] \rightarrow CH_3CHO + H_2O$

ethanol ethanal

Exam tip

When writing equations for the combustion of alcohols many students forget to include the O in the alcohol and it is very common to see the equation for ethanol as:

$$C_2H_5OH + 3\tfrac{1}{2}O_2 \rightarrow 2CO_2 + 3H_2O$$

rather than:

$$C_2H_5OH + 3O_2 \rightarrow 2CO_2 + 3H_2O$$

Oxidation of secondary alcohols to ketones

$CH_3CH(OH)CH_3 + [O] \rightarrow CH_3COCH_3 + H_2O$
propan-2-ol propan-2-one

$CH_3CH_2CH(OH)CH_3 + [O] \rightarrow CH_3CH_2COCH_3 + H_2O$
butan-2-ol butanone

Oxidation of primary alcohols to carboxylic acids

$CH_3OH + 2[O] \rightarrow HCOOH + H_2O$
methanol methanoic acid

$CH_3CH_2OH + 2[O] \rightarrow CH_3COOH + H_2O$
ethanol ethanoic acid

Exam tip

Oxidation of primary alcohols to form carboxylic acids often leads to incorrect equations with the most common wrong response being:

$CH_3CH_2OH + [O] \rightarrow CH_3COOH + H_2$

Remember that oxidation reactions *always* produce water as a product. The correct equation is:

$CH_3CH_2OH + 2[O] \rightarrow CH_3COOH + H_2O$

Substitution of alcohols with halide ions

Alcohols undergo a substitution reaction with halide ions in the presence of an acid such as sulfuric acid:

$$C_2H_5{-}OH + Br^- + H^+ \xrightarrow{NaBr/H_2SO_4} C_2H_5{-}Br + H_2O$$

Haloalkanes

Haloalkanes are compounds in which one or more of the hydrogen atoms of an alkane have been replaced by a halogen atom. If one hydrogen has been replaced, the general formula is $C_nH_{2n+1}X$ (where X = F, Cl, Br or I).

Like alcohols, haloalkanes can be subdivided into primary, secondary and tertiary.

The carbon–halogen bond is a polar bond and results in the carbon being susceptible to attack by a nucleophile. A nucleophile is defined as a lone pair (of electrons) donor.

Nucleophilic substitution

As with any mechanism, when describing **nucleophilic** substitution reactions, you should include curly arrows and any relevant dipoles and lone pairs of electrons.

A **nucleophile** is an electron pair donor.

Hydrolysis

- Reagents: NaOH or KOH (an alkali is required)
- Conditions: the solvent used must be water and the reaction mixture must be heated under reflux
- Balanced equation:

$CH_3CH_2Br + NaOH \rightarrow CH_3CH_2OH + NaBr$

The hydrolysis reaction (Figure 40) can be monitored by adding an aqueous ethanolic silver nitrate solution. The substituted halide ion reacts with the Ag^+ ion, producing precipitates of one of the following:

- white AgCl(s)
- cream AgBr(s)
- yellow AgI(s)

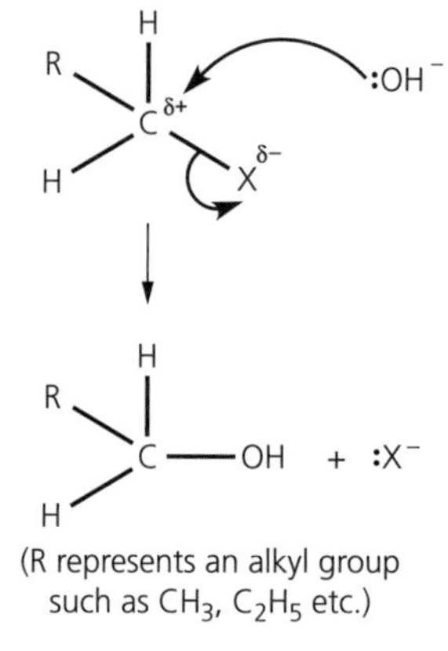

(R represents an alkyl group such as CH_3, C_2H_5 etc.)

Figure 40

Rate of hydrolysis

When equal amounts of 1-chlorobutane, 1-bromobutane and 1-iodobutane are reacted, under identical conditions, with a hot aqueous ethanolic solution of silver nitrate, 1-iodobutane reacts the fastest and 1-chlorobutane the slowest. This can be explained by comparing the carbon–halogen bond enthalpies (Table 14).

Bond	C–F	C–Cl	C–Br	C–I
Bond enthalpy/kJ mol^{-1}	+467	+340	+280	+240

Table 14

Clearly, the C–I bond is the weakest and therefore requires least energy to break it. The C–F is so strong that it rarely, if ever, undergoes hydrolysis.

Environmental concerns from use of organohalogen compounds

Haloalkanes are extremely useful compounds. They are used in the preparation of a wide range of products, including pharmaceuticals (such as ibuprofen); polymers such as PVC, which is made from $CH_2{=}CHCl$; and PTFE which is made from $F_2C{=}CF_2$.

CFCs

CFCs such as dichlorodifluoromethane, CCl_2F_2, and trichlorofluoromethane, CCl_3F, were developed to be used in air conditioning, refrigeration units and aerosols, as well as being used as blowing agents in the production of foamed polymers such as expanded polystyrene. They are suitable for these purposes because they are unreactive, non-flammable and non-toxic, as well as being liquids of low volatility that can be readily evaporated and re-condensed. However, it is also these properties that make them so persistent in the atmosphere. Our knowledge at the time of their introduction did not extend to understanding the dangerous effect they would ultimately have in the stratosphere.

CFCs are blamed for the depletion of the ozone layer. It is thought that when CFCs reach the upper atmosphere they undergo photodissociation and generate chlorine radicals, Cl•. These are extremely reactive and react with ozone (O_3) in the presence of ultraviolet light. The reaction is similar to the reaction between chlorine radicals and alkanes.

Initiation $CCl_2F_2 \rightarrow \bullet CClF_2 + Cl\bullet$

The chlorine radical is then involved in the propagation steps:

Propagation 1 $Cl\bullet + O_3 \rightarrow ClO\bullet + O_2$

Propagation 2 $ClO\bullet + O \rightarrow Cl\bullet + O_2$

The Cl• is regenerated in the second propagation step and can then go on to react with other ozone molecules.

Termination any two radicals reacting together

The exhaust gases from aeroplanes contain nitrogen monoxide, •NO, which is a radical and also reacts with ozone:

Propagation 1 $\bullet NO + O_3 \rightarrow \bullet NO_2 + O_2$

Propagation 2 $\bullet NO_2 + O \rightarrow \bullet NO + O_2$

Knowledge check 29

a Write an equation for:
- **i** the complete combustion of propan-1-ol
- **ii** the dehydration of propan-1-ol
- **iii** the oxidation of propan-1-ol under reflux
- **iv** the oxidation of propan-1-ol under distillation
- **v** the reaction of propan-1-ol with $NaBr/H_2SO_4$

b Explain what is meant by the terms *reflux* and *distillation*.

Organic synthesis

Practical skills

An organic synthesis might be planned, carried out and the product analysed as follows:

Stage 1: planning. Devise a series of reactions that will enable you to make the 'target molecule' from a readily available reagent. Carry out a risk assessment and identify any potential hazards.

Stage 2: carry out the reaction. Decide on:
- the apparatus you will need
- suitable quantities

Stage 3: separation of product.
- Solids by filtration
- Liquids by solvent extraction or distillation

Stage 4: purification of product. The product is usually contaminated with a mixture of unreacted reagents and by-products.
- Solid products are purified by recrystallisation.
- Liquids are purified by fractional distillation

Stage 5: measuring percentage yield. Compare the actual yield with the theoretical yield.

Stage 6: identification of product. Functional groups can be identified by simple chemical tests known as 'wet tests'. Individual compounds can be identified by melting or boiling points. Precise identification involves a combination of analysis including MS, IR and NMR (studied in year 2).

Synthetic routes

Detailed knowledge of the properties and reactions of a limited number of functional groups enables the preparation of a wide variety of organic compounds.

Table 15 summarises the reactions of these groups.

Functional group	Reagent	Product functional group
Alkane	Halogen	Haloalkane
Alkene	Hydrogen halides	Haloalkane
	Halogens	Dihaloalkanes
	Steam	Alcohol
	Hydrogen	Alkane
Alcohols	$H^+/Cr_2O_7^{2-}$	Aldehyde, ketone or carboxylic acid
	Hot concentrated H_2SO_4	Alkene
	NaBr in presence of H_2SO_4	Haloalkane
	Carboxylic acids*	Ester*
Haloalkane	NaOH(aq)	Alcohol
	NH_3(ethanol) *	Amine*
	Cyanide, $^-C{\equiv}N$ *	Nitrile*

* Only likely to be tested in the second year of the A-level course.

Table 15

Example

Preparation of butanone starting from but-1-ene.

In this case, the 'target molecule' is butanone and the 'starting molecule', but-1-ene, is an alkene.

Step 1: start with the target molecule and identify the compounds that could readily be converted directly into the target – concentrate on the functional group.

Butanone is a ketone which can be made from the oxidation of a secondary alcohol, butan-2-ol.

Step 2: look at your starting molecule, but-2-ene, what reactions of alkenes do you know?

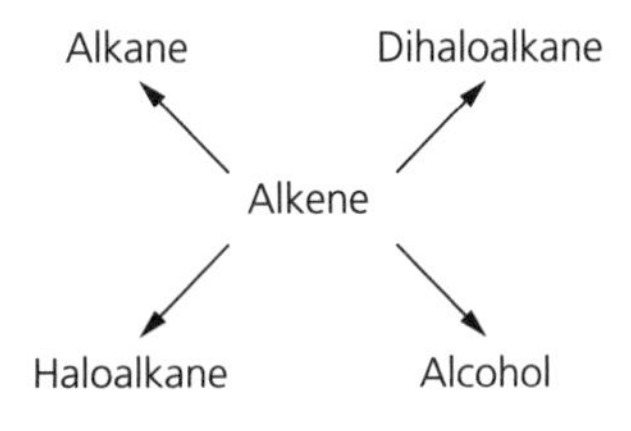

You should now see a possible two-stage synthetic route from your starting molecule to the target molecule. In this case, the route can go via the alcohol.

$$CH_3CH{=}CHCH_3 \rightarrow CH_3CH(OH)CH_2CH_3 \rightarrow CH_3COCH_2CH_3$$

starting molecule   intermediate molecule   target molecule

You need to know the reagents and conditions for each step:

$$H_3C{-}CH{=}CH{-}CH_3 \xrightarrow[\text{H}^+ \text{ catalyst, high temperature and pressure}]{\text{steam}} H_3C{-}CH(OH){-}CH_2{-}CH_3 \xrightarrow[\text{heat under reflux}]{H^+/Cr_2O_7^{2-}} H_3C{-}CO{-}CH_2{-}CH_3$$

You may have to write equations for each step:

Step 1 $CH_3CH{=}CHCH_3 + H_2O \rightarrow CH_3CH(OH)CH_2CH_3$

Step 2 $CH_3CH(OH)CH_2CH_3 + [O] \rightarrow CH_3COCH_2CH_3 + H_2O$

Chemists normally seek a synthetic route that has the least number of stages and which therefore produces a higher yield of the product. It is rare for any one reaction to be 100% efficient. Normally the percentage yield is significantly below the theoretical yield.

Knowledge check 30

Devise a two-stage synthesis for converting:

a ethane to ethanol

b ethanol into ethane

State the reagents and conditions needed for each conversion.

Modern analytical techniques

Infrared spectroscopy

Infrared spectra can be used to identify key absorptions of the alcohol, carbonyl, carboxylic acid and amine functional groups (Table 16). (Datasheets will be supplied in all examinations, so there is no need to learn these values.)

Bond	Location	Wavenumber/cm^{-1}
C–O**	Alcohols, esters*, carboxylic acids	1000–1300
C=O	Aldehydes, ketones, carboxylic acids, esters*, amides*	1630–1820
C=C	Alkenes	1620–1680
C–H	Any organic compound with a C–H bond	2850–3100
O–H	Carboxylic acids	2500–3100 (very broad)
O–H	Alcohols, phenols*	3200–3600 (broad)
N–H	Amines, amides*	3300–3500

* Functional groups most likely to appear only in the second year of the course
** Difficult to detect as they appear in the fingerprint region where there are very many peaks

Table 16

Infrared spectroscopy is used to distinguish between alcohols and their oxidation products: aldehydes, ketones and carboxylic acids.

For an alcohol you need to identify *one* peak (Figure 41):

- 3200–3600 cm^{-1}: all alcohols contain an O–H bond, which is a broad peak and should not be confused with the small, sharp peaks due to C–H bonds
- 1000–1300 cm^{-1}: all alcohols also contain a C–O bond but they are in the fingerprint region and are difficult to assign

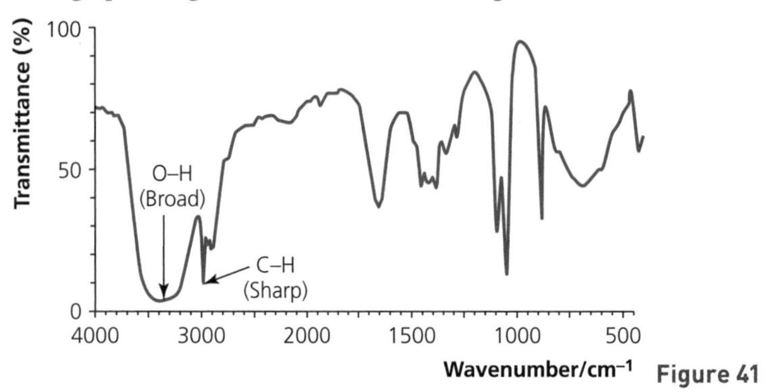

Figure 41

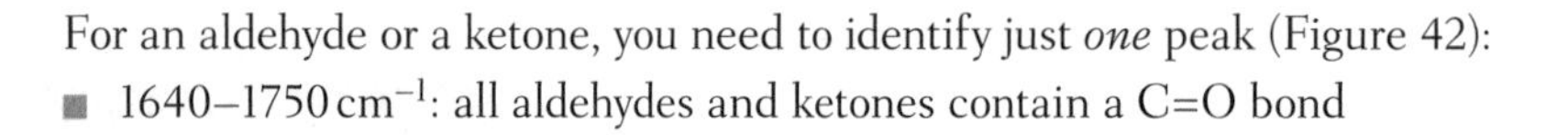

For an aldehyde or a ketone, you need to identify just *one* peak (Figure 42):

- 1640–1750 cm^{-1}: all aldehydes and ketones contain a C=O bond

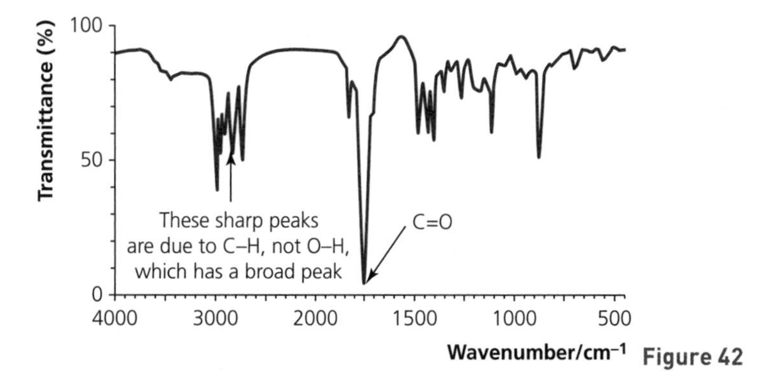

Figure 42

For a carboxylic acid you need to identify *two* peaks (Figure 43):

- 1630–1820 cm^{-1}: all carboxylic acids contain a C=O bond
- 2500–3300 cm^{-1}: all carboxylic acids contain an O–H bond, which is very broad
- 1000–1300 cm^{-1}: all carboxylic acids also contain a C–O bond but they are in the fingerprint region and are difficult to assign

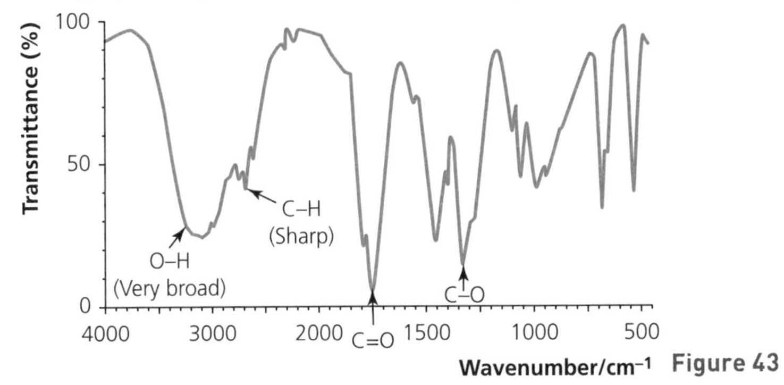

Figure 43

Nowadays, modern analytical techniques have replaced 'wet-tests' or chemical tests. In 1967, when the breathalyser was first introduced into the UK, alcohol (ethanol) was detected using acidified dichromate crystals that changed colour from orange to green. Today this 'wet-test' has been replaced by analysis using infrared spectroscopy.

Infrared spectroscopy is a powerful tool in identifying a particular functional group, but when identifying a specific chemical it is usually used in conjunction with other analytical techniques, such as mass spectrometry, chromatography and nuclear magnetic resonance (NMR) spectroscopy. Chromatography and NMR will be covered in the second year of the course. In year 1 exams you are expected to be able to link together information obtained from an infrared spectrum and a mass spectrum.

Mass spectrometry

Evidence for the existence of isotopes is obtained when using a mass spectrometer. A gaseous sample of the substance is bombarded with high-energy electrons to create positive ions. It is important to remember that the mass spectrometer analyses positive ions only. The positive ions are generated by the interaction of high-energy electrons with the gaseous sample. The high-energy electrons remove an electron from the sample. This can be represented by the following equation, where the sample is X.

$$e^- + X(g) \longrightarrow X^+(g) + 2e^-$$

High-energy electron; Positive ion

The positive ions are passed through a magnetic field, where they are deflected. The amount of deflection is dependent on their mass. A detector is calibrated to record the degree of deflection and to interpret this in terms of the mass of the ions.

For an element, a typical printout from the detector looks like a 'stick-diagram', with each stick representing an ion (isotope). The bigger the 'stick', the more abundant is the ion.

The mass spectrum for boron is shown in Figure 44. The *x*-axis is labelled '*m/z*'. This means mass/charge, but since the charge is 1+ it is effectively the relative mass of the ions

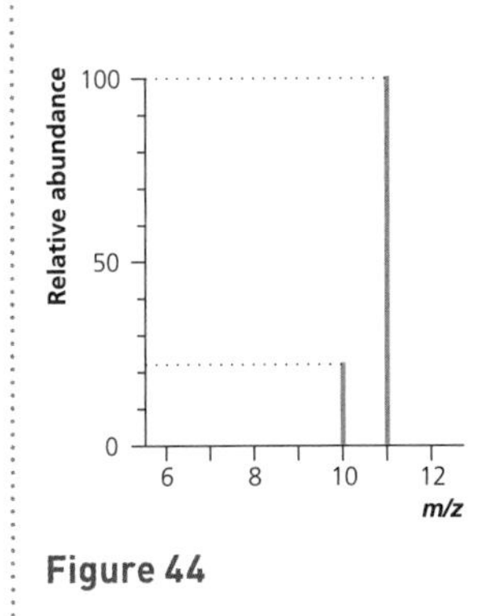

Figure 44

that is recorded. It is calibrated so that the isotope carbon-12 is given a value of exactly 12. This is the standard that is universally used for the comparison of the masses of atoms.

The mass spectrum of boron shows two lines, indicating that there are two isotopes with masses of 10 and 11. The relative abundance of these two isotopes is shown by the heights of the lines — they are in the ratio of 23 (boron-10) to 100 (boron-11).

$$\text{relative atomic mass} = (10 \times 23) + \frac{11 \times 100}{123} = 10.8$$

The mass spectrometer is able to work with very small samples of gaseous molecules, and can be operated remotely. This has made it an extremely useful piece of equipment for analysing samples in remote locations. The equipment on board the Mars space probe included a mass spectrometer that was able to analyse gaseous samples and return the results to Earth.

The mass spectrometer can also analyse compounds, although the number of lines obtained is usually quite large. This is because bombardment by electrons causes the molecule to break up, with each of the fragments obtained registering on the detector. This can be an advantage, as it is sometimes possible to obtain details of the structure of the molecule as well as its overall molecular mass.

Figure 45 is the mass spectrum of propane (C_3H_8).

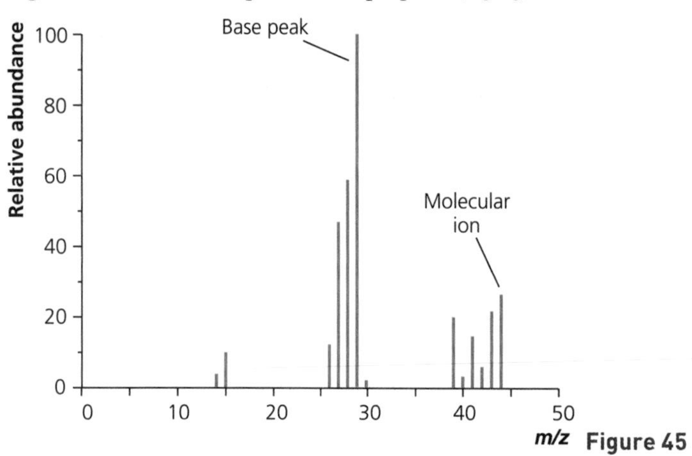

Figure 45

Propane has a relative molecular mass of 44 and, as expected, the peak furthest to the right of the spectrum represents the ion $C_3H_8^+$. This is called the **molecular ion peak** (Figure 46).

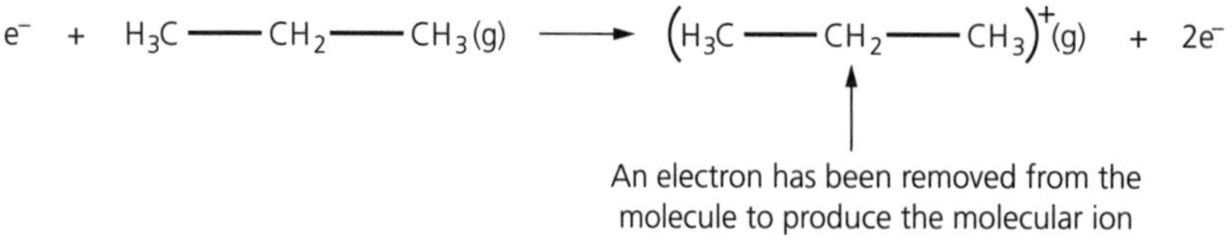

$$e^- + H_3C{-}CH_2{-}CH_3(g) \longrightarrow (H_3C{-}CH_2{-}CH_3)^+(g) + 2e^-$$

An electron has been removed from the molecule to produce the molecular ion

Figure 46

However, bombardment by electrons also creates ions of fragments of the molecule. These ions will also be registered on the printout. The peak at 29 occurs because a CH_3 unit has been broken from the $CH_3CH_2CH_3$ chain, and thus the ion $CH_3CH_2^+$ has been detected (Figure 47).

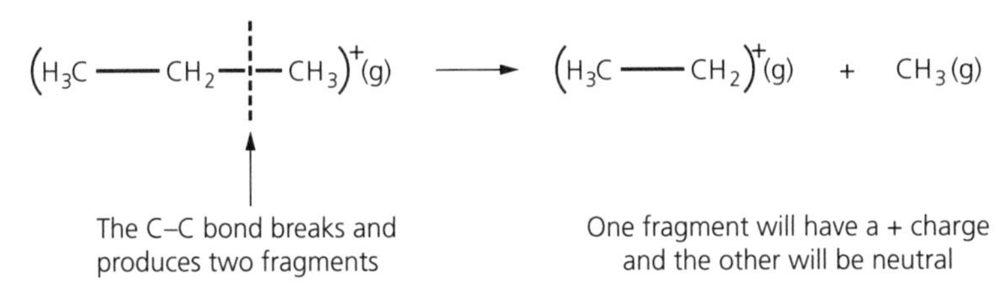

Figure 47

Similarly, the peak at 15 represents a CH_3^+ ion (Figure 48) and it is possible to suggest the identity of all other peaks in the spectrum.

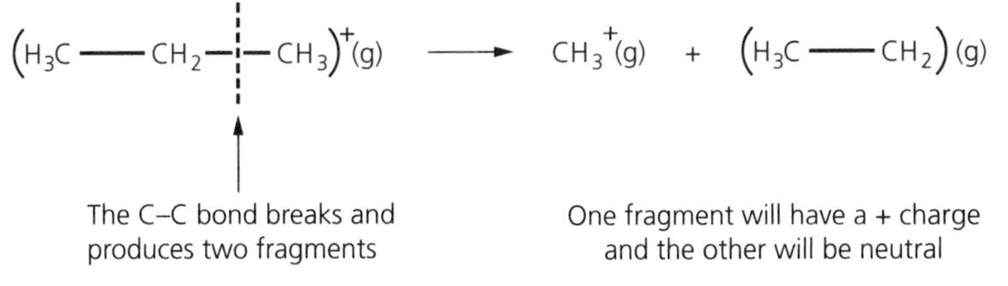

Figure 48

Fragmentation leads to a large number of peaks, and the pattern of these is often said to be a fingerprint of the molecule. Like a fingerprint, it can be used in conjunction with a computer containing a spectral database to identify a particular chemical.

Butan-1-ol and butan-2-ol have very similar infrared spectra. Their mass spectra are shown in Figure 49.

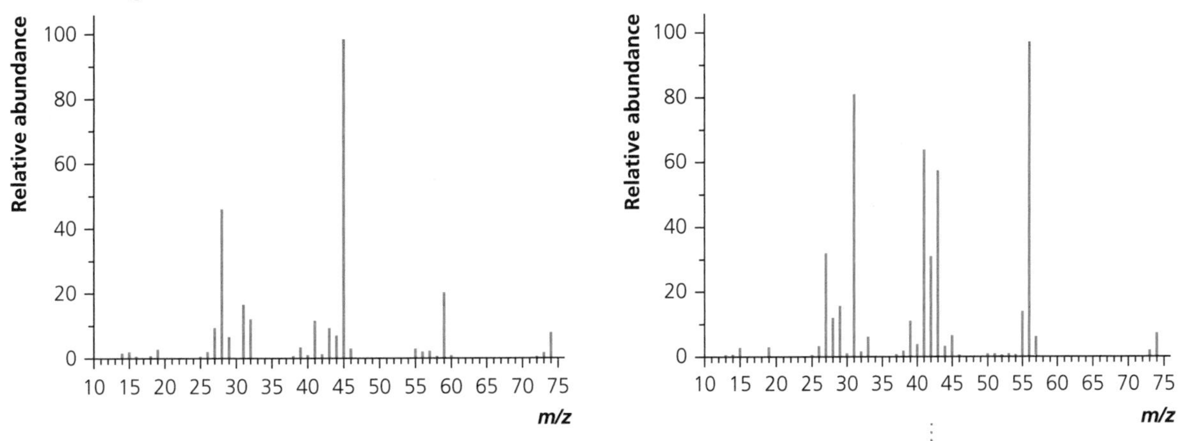

Figure 49 Both spectra have molecular ion peaks at 74, but the fragmentation ions are very different. These can be used to distinguish between the two isomers

Analysis of the fragmentation of molecules can be used to distinguish between structural isomers. Although this may not always be easily apparent at a glance, computers can be used to compare the pattern obtained from a compound with a database of spectra, and this is a widely used research tool.

Knowledge check 31

In the mass spectrum of butan-2-ol suggest the identity of peaks with *m/z* values of: 15, 45, 59 and 74.

Combined techniques

You will be expected to use a combination of techniques to deduce the structure of organic compounds using a range of different analytical techniques. These will include:

- empirical formula calculations
- qualitative analysis (wet tests for functional groups)
- mass spectra
- IR spectra

Knowledge check 32

The mass spectrum and the infrared spectrum of compound A are shown below. Show that the compound is but-3-en-1-ol, $CH_2{=}CHCH_2CH_2OH$.

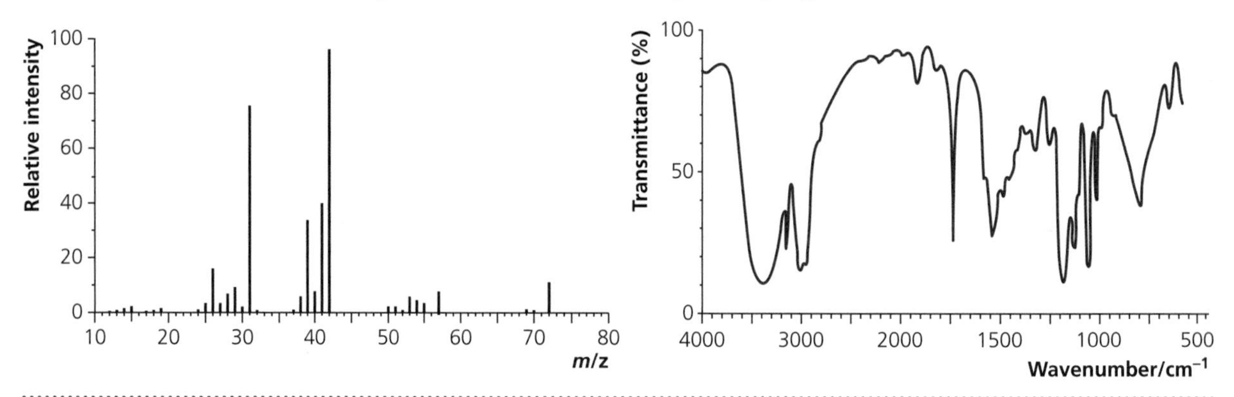

Summary

Having revised **Module 4: Alcohols, haloalkanes and analysis** you should now have an understanding of:

- reactions of alcohols including oxidation, elimination and substitution with halide ions
- hydrolysis of haloalkanes
- nucleophilic substitution mechanism
- CFCs and ozone
- organic synthesis
- how to recognise absorptions due to O–H and C=O bonds in infrared spectra
- how to determine the molar mass of molecules by using the molecular ion peak in mass spectra

Questions & Answers

Approaching the exam

There are four modules in the AS (H032) specification and a further two modules in the A-level (H342) specification. The table summarises the structure and content of the exams for AS and for A-level.

All questions used in this book are relevant to both AS and to A-level examinations unless otherwise stated.

	Component exam	Modules						Total marks/ time	Type of questions
		1	2	3	4	5	6		
AS	Breadth in chemistry	✓	✓	✓	✓			70 marks 1 hour 30 minutes	Multiple choice (20 marks) Short-answer questions (50 marks)
	Depth in chemistry	✓	✓	✓	✓			70 marks 1 hour 30 minutes	Short-answer and extended-response questions (70 marks)
A-level	Periodic table and physical chemistry	✓	✓	✓		✓		100 marks 2 hours 15 minutes	Multiple choice (15 marks) Short-answer and extended-response questions (85 marks)
	Synthesis and analytical techniques	✓	✓		✓		✓	100 marks 2 hours 15 minutes	Multiple choice (15 marks) Short-answer and extended-response questions (85 marks)
	Unified chemistry	✓	✓	✓	✓	✓	✓	70 marks 1 hour 30 minutes	Short-answer and extended-response questions (70 marks)

Question types

Multiple-choice questions will be used in AS (paper 1) and in A-level papers 1 and 2. Multiple-choice questions need to be read carefully and there is often a misconception that these questions have to be done 'in your head'. That is not the case; many multiple-choice questions require you to think and to work things out on paper and space will be provided on the question paper. For each of the questions there are four suggested answers; A, B, C or D. You select your response by putting a cross in a box by the letter of your choice. Multiple-choice questions are machine-marked, and it is essential that you follow the instructions given on the exam paper. The multiple-choice questions in this book will not be 'machine-marked' and therefore do not follow the exam format exactly.

Short-answer questions will appear in all examinations at both AS and at A-level. Read the questions carefully, be aware of the marks awarded for each section — for example if there are 2 marks for a sub-section in a question you will be expected to make two points. The space provided for the answer has been designed to be more than sufficient for a complete response. Do not write in the margins. If you do require additional space, ask for extra paper and explain on the question paper that you have used extra paper for this response. Typical short-answer questions are illustrated in this book by questions 11–22.

Extended-response questions will appear on paper 2 at AS and on all three papers at A-level. Students often think of this type of question as 'essay' questions, but you don't have to use an essay style when answering these questions. Chemists communicate in a variety of ways such as formulae, equations, mechanisms all of which are difficult to put into an essay style. It is perfectly acceptable to use bullet points or to tabulate your response. Extended-response questions require that you plan your answer carefully. Make sure you use the marks allocated as a guidance. If there are 8 marks you must make eight separate points. Check the mark schemes from previous papers and you will see that each mark is allocated to a specific point. Typical extended-response questions are illustrated in this book by questions 23–26.

Terms used in examination questions

You will be asked precise questions in the examinations, so you can save a lot of valuable time (as well as ensuring you score as many marks as possible) by knowing what is expected. Terms most commonly used are explained below.

Define

This is intended literally. Only a formal statement or equivalent paraphrase is required.

Explain

This normally implies that a definition should be given, together with some relevant comment on the significance or context of the term(s) concerned, especially where two or more terms are included in the question. The amount of supplementary comment intended should be determined by the mark allocation.

State

This implies a concise answer with little or no supporting argument.

Describe

This requires you to state in words (using diagrams where appropriate) the main points of the topic. It is often used with reference either to particular phenomena or to particular experiments. In the former instance, the term usually implies that the answer should include reference to (visual) observations associated with the phenomena. The amount of description intended should be interpreted according to the indicated mark value.

Deduce or predict

This implies that you are not expected to produce the required answer by recall but by making a logical connection between other pieces of information. Such information may be wholly given in the question or may depend on answers given in an earlier part of the question. 'Predict' also implies a concise answer, with no supporting statement required.

Outline

This implies brevity, i.e. restricting the answer to essential detail only.

Suggest

This is used in two main contexts. It implies either that there is no unique answer or that you are expected to apply your general knowledge to a 'novel' situation that may not be formally in the specification.

Calculate

This is used when a numerical answer is required. In general, working should be shown.

Sketch

When this is applied to diagrams, it implies that a simple, freehand drawing is acceptable. Nevertheless, care should be taken over proportions, and important details should be labelled clearly.

About this section

This section contains questions similar in style to those you can expect to find in your AS and A-level examinations. Each question in this section identifies the specification topic, the total marks and a suggested time that should be spent writing out the answer.

The limited number of questions means that it is impossible to cover all the topics and question styles, but they should give you a flavour of what to expect. The responses that are shown are real students' answers to the questions. Student A is an A/B-grade student and Student B is a B/C-grade student.

There are several ways of using this section. You could:

- 'hide' the answers to each question and try the question yourself. It needn't be a memory test — use your notes to see if you can make all the necessary points
- check your answers against the students' responses and make an estimate of the likely standard of your response to each question
- take on the role of the examiner and mark each student's response, then check whether you agree with the marks awarded
- check your answers against the comments to see if you can appreciate where you might have lost marks

Comments

Comments on the questions are preceded by the icon ⓔ. They offer tips on what you need to do in order to gain full marks. All student responses are followed by comments, indicated by the icon ⓔ, which highlight where credit is due. In the weaker answers, they also point out areas for improvement; specific problems; and common errors such as lack of clarity, irrelevance, misinterpretation of the question and mistaken meanings of terms.

Multiple-choice questions

Questions in this section are relevant to AS Component 1 and to A-level Components 1 and 2.

Multiple-choice questions similar to the ones below will appear on AS paper 1 and on A-level papers 1 and 2.

Answer the questions that follow and record your answers. Check your answers when you have completed all 10 questions.

Question 1

The melting points of four successive elements in period 3 are listed below.

649°C 660°C 1410°C 290°C

The four elements are:

A Na, Mg, Al, Si **B** Mg, Al, Si, P **C** Al, Si, P, S **D** Si, P, S, Cl

Question 2

Which one of the following equations correctly represents the enthalpy change of formation of ethene from graphite, C(s)?

A $2C(s) + 4H(g) \rightarrow C_2H_4(g)$

B $2C(s) + 2H_2(g) \rightarrow C_2H_4(g)$

C $C(s) + H_2(g) \rightarrow \frac{1}{2} C_2H_4(g)$

D $2C(g) + 2H_2(g) \rightarrow C_2H_4(g)$

Question 3

The equilibrium concentration of the products of the reaction:

$$H_2O(g) + C(s) \rightleftharpoons H_2(g) + CO(g) \qquad \Delta H = +180.5\,kJ$$

can be increased most by:

A raising the temperature only

B raising the temperature and lowering the pressure

C raising the temperature and increasing the pressure

D lowering the temperature and increasing the pressure

Question 4

Which one of the following is *not* readily oxidised?

A hexan-2-ol **B** hexan-3-ol **C** 2-methylpentan-2-ol **D** 3-methylpentan-2-ol

Question 5

At room temperature and pressure, which one of the following is the minimum volume of oxygen required for complete combustion of 16 g of methane?

A $2\,dm^3$ **B** $64\,dm^3$ **C** $24\,dm^3$ **D** $48\,dm^3$

Question 6

Ethanol, ethanal and ethanoic acid can best be separated by:

A filtration **B** separating funnel **C** distillation **D** reflux

Question 7

Haloalkanes such as CH_3CH_2Cl, CH_3CH_2Br and CH_3CH_2I can be hydrolysed but the rate of hydrolysis varies. Which of the following statements is true?

A CH_3CH_2I would be the fastest because the C–halogen bond has the biggest dipole.

B CH_3CH_2Cl would be the fastest because the C–halogen bond has the biggest dipole.

C CH_3CH_2I would be the fastest because it has the longest C–halogen bond.

D CH_3CH_2Cl would be the fastest because it has the longest C–halogen bond.

Use the following key to answer questions 8–10.

A	B	C	D
1, 2 and 3 correct	1 and 2 correct	1 and 3 correct	3 only correct

Question 8

Which of the following is always true for an endothermic reaction?

1 The total enthalpy of the products is greater than the total enthalpy of the reactants.

2 Heat will be absorbed from the surroundings when the reaction takes place.

3 The activation energy for the conversion of the products into the reactants is less than the activation energy for the conversion of the reactants into the products.

Question 9

Compound X, $CH_2C(OH)CH_3$, will:

1 decolorise bromine

2 react with $H^+/Cr_2O_7^{2-}$ and the colour will change from green to orange

3 react with $H^+/Cr_2O_7^{2-}$ and the colour will change from orange to green

Question 10

Which of the following statements about the reactions of 2-methylpropene are correct?

1 It reacts with hydrogen to form a saturated hydrocarbon with empirical formula C_2H_5.

2 When it reacts with steam in the presence of an acid catalyst one of the products obtained is a tertiary alcohol.

3 It has a higher boiling point than but-1-ene.

Answers to questions 1–10

The student answers are *all incorrect* and are used to illustrate the most common incorrect response.

	1	2	3	4	5	6	7	8	9	10
Student answers	A	A	A	B	A	A	B	A	D	A
Correct answers	B	B	B	C	D	C	C	A	C	B

e It should now be apparent that you cannot answer all multiple-choice questions in your head — many require working out on paper.

Q1 In period 3, melting points increase across the period to a maximum with Si, which is then followed by a large drop in melting point.

Q2 Alternative A looks feasible but by definition the elements must be in their natural states.

Q3 If you ignore the state symbol for carbon, alternative A seems to be correct, but the reaction has greater volume on the right-hand side of the equation — C(s) is not a gas so is excluded — and the forward reaction is endothermic.

Q4 Don't be fooled by the name, hexan-3-ol is not a tertiary alcohol. Take your time and sketch the alternatives before deciding which is the correct response.

Q5 Difficult to do without writing the equation first. 16 g of CH_4 is 1 mole of CH_4, which therefore requires 2 mols of O_2, which has a volume of $2 \times 24 = 48\,dm^3$.

Q6 They are all liquids and will have different boiling points, which enable separation by fractional distillation.

Q7 Rate depends on bond enthalpy rather than difference in electronegativity and C–I is the weakest bond; the longer the bond length the weaker the bond.

Q8 It's a good idea to sketch the enthalpy profile and then consider each statement separately.

Q9 Compound X, $CH_2C(OH)CH_3$, contains a C=C double bond but it is hidden. The question would be much easier if it had been written as $CH_2{=}C(OH)CH_3$ rather than $CH_2C(OH)CH_3$.

Q10 Again difficult to do in your head — it helps to sketch out each reaction.

Short-response questions

Questions in this section are relevant to AS Components 1 and 2 and to A-level Components 1, 2 and 3.

Question 11 Ionisation energies

Time allocation: 9–10 minutes

Successive ionisation energies can provide evidence for the electronic configuration of an element. The table below provides data on the successive ionisation energies of oxygen.

Ionisation number	1st	2nd	3rd	4th	5th	6th	7th	8th
Ionisation energy/ $kJ\,mol^{-1}$	1314	3388	5301	7469	10989	13327	71337	84080
Log (ionisation energy)	3.1	3.5	3.7	3.9	4.0	4.1	4.8	4.9

(a) **(i)** Plot log (ionisation energy) against ionisation number. Explain how these data provide evidence for the electronic configuration of oxygen. (3 marks)

(ii) Explain why the successive ionisation energies of oxygen increase. (2 marks)

(iii) Write an equation to represent the third ionisation energy of oxygen. (2 marks)

(b) The first five ionisation energies of element X are shown the table below. State, with a reason, the group in which you would expect to find element X. (2 marks)

Ionisation number	1st	2nd	3rd	4th	5th
Ionisation energy/$kJ\,mol^{-1}$	578	1817	2745	11578	14831

(c) Use the grid below to sketch the plot of log (ionisation energy) against ionisation number for phosphorus, $_{15}P$. (3 marks)

Total: 12 marks

ⓔ When asked to 'plot' a graph it is important to use the graph paper fully and to clearly label the axes.

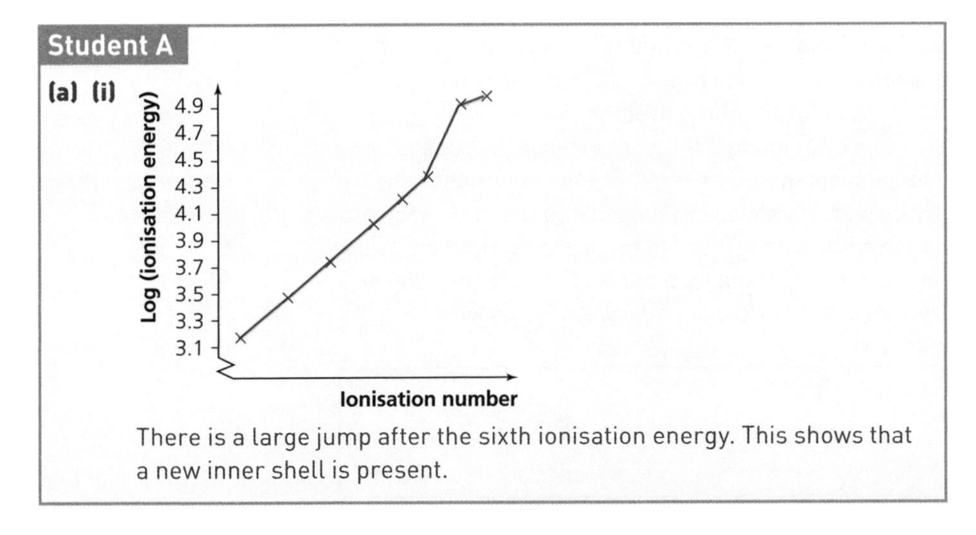

Student B

(a) (i)

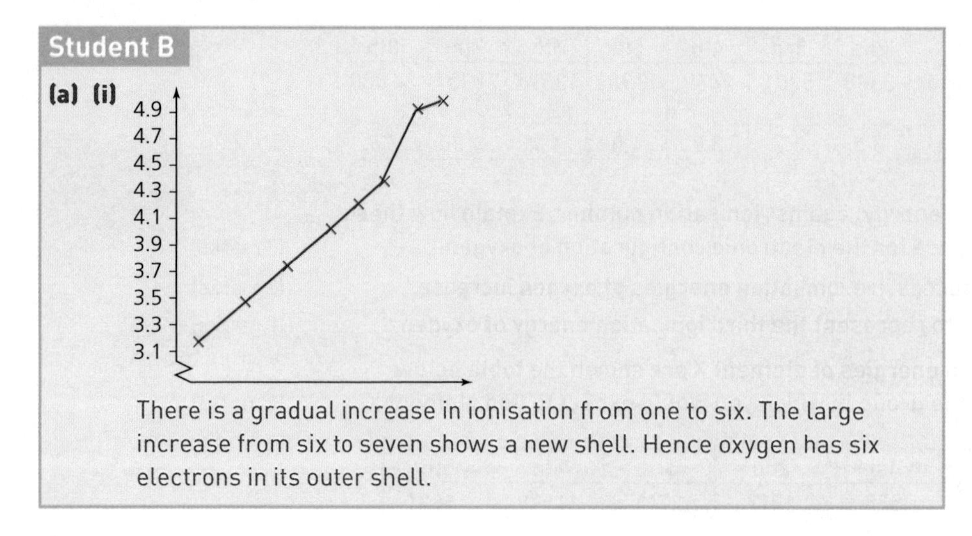

There is a gradual increase in ionisation from one to six. The large increase from six to seven shows a new shell. Hence oxygen has six electrons in its outer shell.

e Student A earns 2 marks for the graph, but Student B loses a mark for not labelling the axes. Both are awarded a mark for relating the large increase in ionisation energy, which occurs between the sixth and the seventh ionisation, to evidence for a new shell.

Student A

(a) (ii) When electrons are removed, the ionic radii decrease due to an increase in the size of the nucleus.

Student B

(a) (ii) Each time the nucleus gets bigger, it is harder to remove the next electron.

e Both students have misunderstood the process. The nucleus remains the same throughout. Oxygen always has eight protons, and this never changes. What does change is the proton to electron ratio. Each time an electron is removed, the proton to electron ratio changes in favour of the protons. Student A picks up 1 mark out of the 2 marks awarded, by pointing out that the radius decreases, but Student B scores zero. The most important factor is the change in radius — the loss of electrons results in an ion with a smaller radius, making the next electron more difficult to remove because it is located closer to the nucleus. Both students seem to understand what is happening, but both lose marks because they do not express themselves precisely. The examiner cannot *interpret* what is written.

Student A

(a) (iii) $O^{2+}(g) \rightarrow O^{3+}(g) + e^-$

Student B

(a) (iii) $O_2(g) \rightarrow O_2^{3+}(g) + 3e^-$

ⓔ Student A is awarded both marks, but Student B needs to carefully revise the basic definitions, having made two fundamental errors:

- ionisation energy relates to *atoms* not molecules
- ionisation energy relates to the removal of *one electron at a time* (not three)

Ionisation energy is defined as the removal of one electron from each atom/ion in one mole of gaseous atoms/ions. The general equation is $X^{(n-1)+}(g) \rightarrow X^{n+}(g) + e^-$, where n = 1, 2, 3 etc.

Student A

(b) Group 3, because the biggest increase in ionisation energy is between the third and fourth ionisation energies. This shows that the fourth electron must be in an inner shell much closer to the nucleus.

Student B

(b) 3

ⓔ Student A gets both marks and gives a good, reasoned answer. Student B also has the correct answer, but only scores 1 of the 2 marks available. Student B needs to read the question carefully and to look at how the marks are allocated.

Student A

(c)

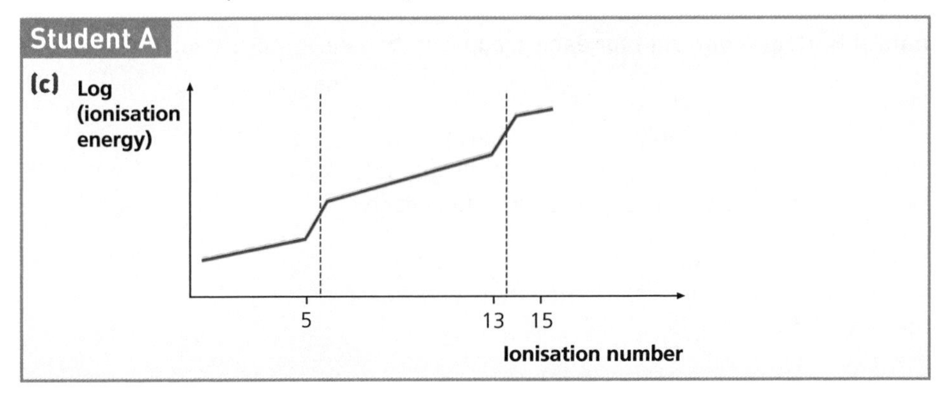

Student B

(c)

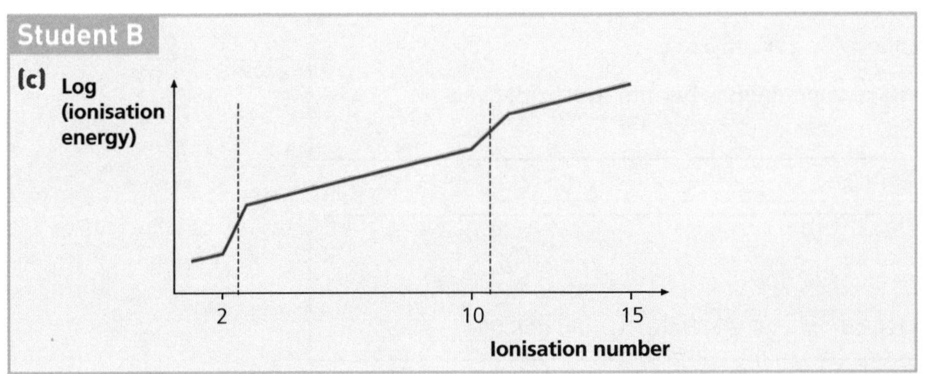

Ⓔ Marks are awarded for recognising that because phosphorus is in group 5, there will be large increases in the ionisation energy after the fifth and the thirteenth (5 + 8) electrons are removed. Student A gains all 3 marks, but Student B loses 2 marks by drawing the ionisation energies back-to-front. This is a mistake made by many students, who tend to think along the lines of $_{15}P$, and therefore use the electronic configuration 2,8,5 and draw a sketch showing that the inner electrons are removed first. It is worthwhile reading through Student B's answer again, listing ways in which additional marks could have been gained with a little more care.

Ⓔ Looking at each answer in isolation, there seems to be little difference in standard between the two students. Student A is more careful and appears to have learnt the basic principles. Student B needs to improve his/her basic knowledge. Overall, Student A scores 11 marks out of 12 (grade A) and Student B scores just 4, which is grade E/U.

Question 12 Group 2

Time allocation: 9–10 minutes

(a) **Barium is a group 2 element. It reacts with oxygen to form compound A, and with water to form compound B and gas C. With the aid of suitable equations, identify A, B and C.** (5 marks)

(b) **(i)** **Calculate the minimum volume of 0.05 mol dm^{-3} HCl(aq) that would be needed to react separately with 1.00 g of barium carbonate and 1.00 g calcium carbonate. Show all your working.** (5 marks)

(ii) **Explain why the volume of HCl(aq) is different for each group 2 carbonate.** (2 marks)

Total: 12 marks

Ⓔ The commands in each section are clear as should be the allocation of the marks. In part (a) the identification of A, B and C (each scores 1 mark) and the command word 'identify' allows you to either name or give the formula of each. Each equation also scores 1 mark.

In (b) the calculation is open-ended and involves several steps. Plan how you are going to do the calculation before you start.

Student A

(a) $Ba + \frac{1}{2}O_2 \rightarrow BaO$; compound A is barium oxide.

$Ba + 2H_2O \rightarrow Ba(OH)_2 + H_2$; compound B is barium hydroxide, and gas C is hydrogen.

Student B

(a) $2Ba + O_2 \rightarrow 2BaO$; compound A is BaO.

$Ba + 2H_2O \rightarrow Ba(OH)_2 + H_2$; compound B is $Ba(OH)_2$, and gas C is H_2.

e These are excellent answers from both students, and they will score maximum marks. When the question asks you to identify a substance, it can be identified either by name or by formula, as long as it is unambiguous. It is safer to name the compounds, as mistakes are often made when quoting formulae. Equations have to be balanced, and it is acceptable to balance them by using fractions if appropriate.

Student A

(b) (i)

Compound	$BaCO_3$	$CaCO_3$
Molar mass	137.3 + 12 + 48 = 197.3 g mol^{-1}	40.1 + 12 + 48 = 100.1 g mol^{-1}
Moles of $BaCO_3$	1/197.3 = 0.005068 = 0.005 mol	1/100.1 = 0.00999 = 0.01 mol
Moles of HCl	0.005 mol	0.01 mol
Volume of HCl	n/c = 0.005/0.05 = 0.1 dm^3 = 100 cm^3	n/c = 0.01/0.05 = 0.2 dm^3 = 200 cm^3

Student B

(b) (i) 202.7 cm^3 and 399.6 cm^3

e Student B gets all 5 marks, but displays an appalling examination technique. If the numerical values had been incorrect, Student B would have lost all 5 marks. Student A's approach is much more sensible. Each step has been clearly shown. Student A has not used the mole ratio when $BaCO_3$ reacts with HCl (the ratio is 1:2). This one error in the calculation would therefore only lose 1 mark. Student A's technique is not good because he/she has rounded up in the middle of the calculation, and it is best to keep the numbers in the calculator until the final step and then round to an appropriate number of significant figures. The correct calculator values are 101.3684744 and 199.8001998, which to 3 significant figures are 101 cm^3 and 200 cm^3 respectively. This would cost Student A a second mark. Student A scores 3 marks.

Student A

(b) (ii) The volumes are different because the chemicals are different.

Student B

(b) (ii) It's the number of moles that's important not the mass.

e Student B gets both marks and Student A scores no marks. In (b)(i) both students should have written equations for the two reactions and the mole ratio of the reactants would have been obvious. Failing to do this has probably cost Student A four marks. Student B has done brilliantly but has shown awful examination technique. On another day an arithmetic error could have cost as many as 7 marks.

e **Overall, Student A scores 8 marks out of 12 and Student B full marks.**

Question 13 The halogens

Time allocation: 11–12 minutes

(a) Chlorine bleach is made by the reaction of chlorine with aqueous sodium hydroxide. In this reaction the oxidation number of chlorine changes and it is said to undergo disproportionation.

$Cl_2(g) + 2NaOH(aq) \rightarrow NaClO(aq) + NaCl(aq) + H_2O(l)$

(i) Determine the oxidation number of chlorine in Cl_2, NaClO and NaCl. (3 marks)

(ii) State what is meant by the term 'disproportionation'. (1 mark)

(iii) The bleaching agent is the ClO^- ion. In the presence of sunlight, this ion decomposes to release oxygen gas. Construct an equation for this reaction. (2 marks)

(iv) If chlorine is reacted with concentrated sodium hydroxide at high temperature $NaClO_3$ is formed rather than NaClO. Construct a balanced equation for this reaction. (2 marks)

(b) The sea contains a low concentration of bromide ions. Bromine can be extracted from seawater by first concentrating the seawater and then bubbling chlorine through this solution.

(i) Suggest how seawater could be concentrated. (1 mark)

(ii) The chlorine oxidises bromide ions to bromine. Construct a balanced ionic equation for this reaction. (1 mark)

(c) Vinyl chloride is a compound of chlorine, carbon and hydrogen. It is used to make polyvinylchloride (PVC). Vinyl chloride has the following percentage composition by mass: chlorine 56.8%; carbon 38.4%; hydrogen 4.8%.

(i) Show that the empirical formula of vinyl chloride is C_2H_3Cl. Show your working. (2 marks)

(ii) The molecular formula of vinyl chloride is the same as its empirical formula. Draw a possible structure, including bond angles, for a molecule of vinyl chloride. (2 marks)

Total: 14 marks

e The command word 'construct' in (a)(iv) indicates that you are not expected to know this equation but that you should use the information in the question to 'build' a new equation. The command word 'suggest' in part (b)(i) indicates that you have to apply your knowledge and apply it to a problem that may not be explicitly on the specification. To 'concentrate' seawater the obvious answer is to remove the water by some method such as evaporation or distillation.

Student A

(a) (i) $Cl_2(g) + 2NaOH(aq) \rightarrow NaClO(aq) + NaCl(aq) + H_2O(l)$

0 +1 −1

Student B

(a) (i) $Cl_2(g) + 2NaOH(aq) \rightarrow NaClO(aq) + NaCl(aq) + H_2O(l)$

0 1 1

e Student A scores all 3 marks and Student B scores 2 marks. Oxidation number has a sign as well as a value. It is always necessary to include the minus sign for negative oxidation numbers. If the oxidation number is positive, the '+' sign should be written, but the examiner will assume that the number is positive if no sign is given.

Student A

(a) (ii) Chlorine is not proportional.

Student B

(a) (ii) Chlorine will displace bromine and iodine.

e Neither student scores the mark. The question command word 'State...' indicates that this is recall and that you should have learnt the definition. In Module 3, disproportionation is defined as 'a reaction in which an element is simultaneously oxidised and reduced'.

Student A

(a) (iii) $ClO^- \rightarrow Cl^- + \frac{1}{2}O_2$

Student B

(a) (iii) $2ClO^- \rightarrow 2Cl + O_2$

e Student A scores both marks, while Student B loses a mark by not balancing the equation for charge. There is a net charge of 2– on the left-hand side of the equation, so there must also be a net charge of 2– on the right-hand side. The 2Cl should be $2Cl^-$.

Student A

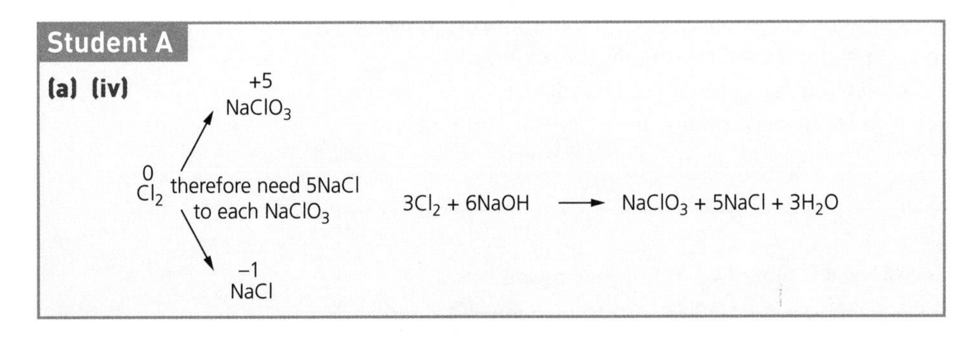

Student B

(a) (iv) $3Cl_2 + 3NaOH \rightarrow NaClO_3 + 2NaCl + 3HCl$

Student A scores both marks and gives a brilliant answer to a difficult question. Student B has tried hard and the equation is balanced but there is a fundamental mistake. It is not possible to have a base (NaOH) as a reagent and an acid (HCl) as a product. They react to produce water.

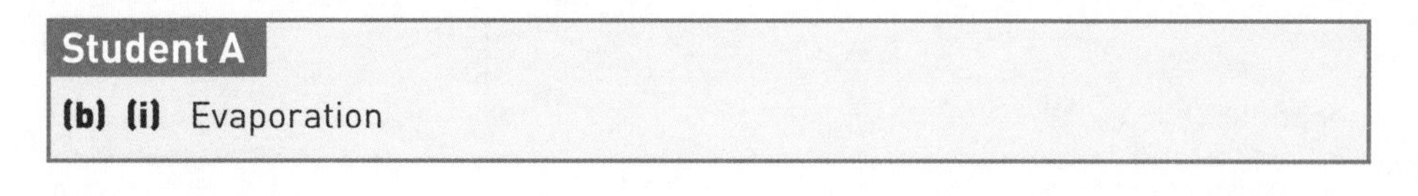

Student A

(b) (i) Evaporation

Student B

(b) (i) Heat

Both students score the mark because both methods would work.

Student A

(b) (ii) $Cl_2 + 2Br^- \rightarrow Br_2 + 2Cl$

Student B

(b) (ii) $Cl_2 + Br_2 \rightarrow 2Cl^- + 2Br^-$

Neither student scores the mark. Student A has made the same mistake that Student B made in part (a)(iii) and has forgotten to balance the net charges. Student B has completely confused bromine and bromide, chlorine and chloride. The instructions in the question tell you exactly what is happening but you have to read the question carefully.

Questions on displacement reactions should be straightforward recall of knowledge. However, students seem to find ionic equations difficult and even able students get them wrong. Many students are confused by the names 'bromide' and 'bromine'. It is worth spending some time ensuring that you know and understand these reactions and that you recognise that a word ending in '-ide' is an ion and has a negative charge and a word ending in '-ine' is an atom or a diatomic molecule.

Student A

(c) (i) Divide the percentage of each element by its relative atomic mass:
chlorine (56.8/35.5) = 1.6, carbon (38.4/12.0) = 3.2, hydrogen = (4.8/1.0) = 4.8
Ratio is 1.6:3.2:4.8
Divide by the smallest: 1.6/1.6:3.2/1.6:4.8/1.6
Therefore, the ratio is 1:2:3, which is equivalent to ClC_2H_3.
So the empirical formula is C_2H_3Cl.

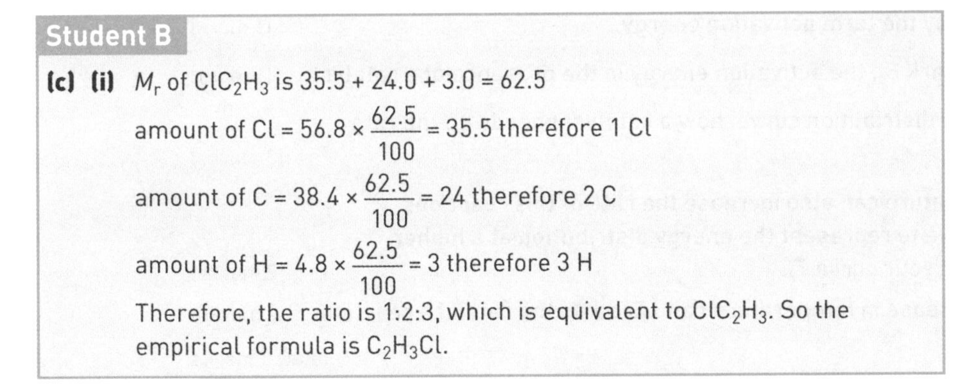

Student B

(c) (i) M_r of ClC_2H_3 is 35.5 + 24.0 + 3.0 = 62.5

amount of Cl = $56.8 \times \frac{62.5}{100}$ = 35.5 therefore 1 Cl

amount of C = $38.4 \times \frac{62.5}{100}$ = 24 therefore 2 C

amount of H = $4.8 \times \frac{62.5}{100}$ = 3 therefore 3 H

Therefore, the ratio is 1:2:3, which is equivalent to ClC_2H_3. So the empirical formula is C_2H_3Cl.

ⓔ Each student scores 2 marks. Students A and B use different methods to calculate the empirical formula, but both are valid and hence earn the marks.

ⓔ A correct C=C double bond automatically earns 1 mark. The second mark is for the rest of the sketch and the bond angle. Each student scores 1 mark. Student A loses a mark because there is no C=C double bond; Student B loses a mark because the shape is not correct and a bond angle has not been included.

ⓔ **Overall, Student A scores 11 out of 14 marks; Student B scores 7.**

Question 14 Boltzmann distribution

Time allocation: 6–7 minutes

The figure below shows the energy distribution of reactant molecules at a temperature T_1.

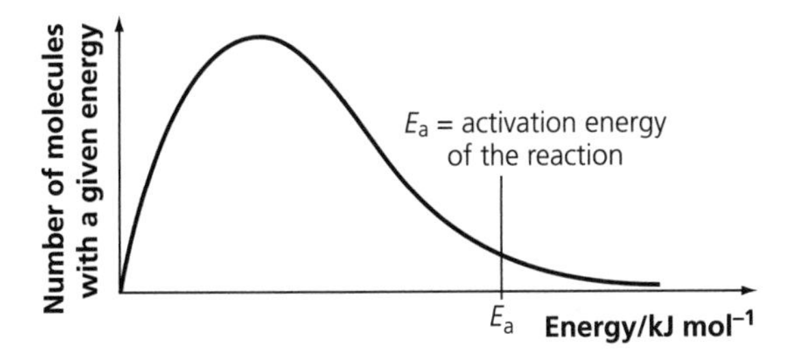

(a) Explain what is meant by the term activation energy. (1 mark)

(b) On the sketch above, mark E_c, the activation energy in the presence of a catalyst. (1 mark)

(c) Explain, in terms of the distribution curve, how a catalyst speeds up the rate of a reaction. (2 marks)

(d) **(i)** Raising the temperature can also increase the rate of this reaction. Draw a second curve to represent the energy distribution at a higher temperature. Label your curve T_2. (2 marks)

(ii) Explain how an increase in temperature can speed up the rate of a reaction. (2 marks)

Total: 8 marks

ⓔ The command word 'explain' in (a), (c) and (d)(ii) infers that reasons are needed to support your statements. The command word 'draw' in (d)(i) is self-explanatory and a diagram is required.

Student A

(a) The activation energy is the minimum energy needed to start a reaction.

Student B

(a) The energy needed for a collision to be successful.

ⓔ Student A gains the mark, but Student B doesn't. The key word missing from Student B's response is *minimum*.

Student A

(b)

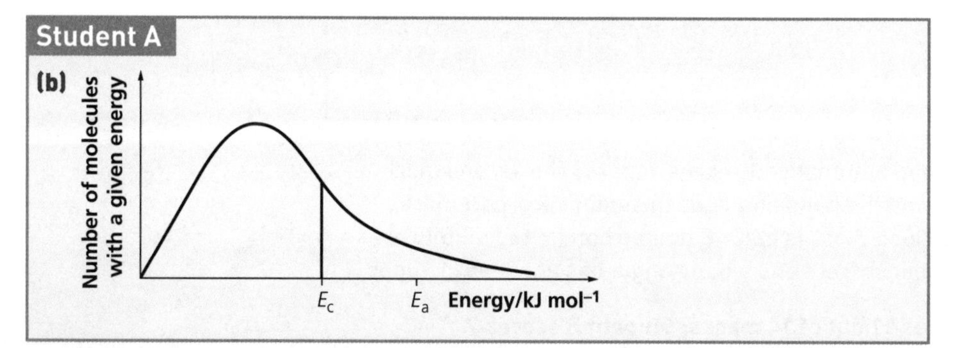

Student B

(b)

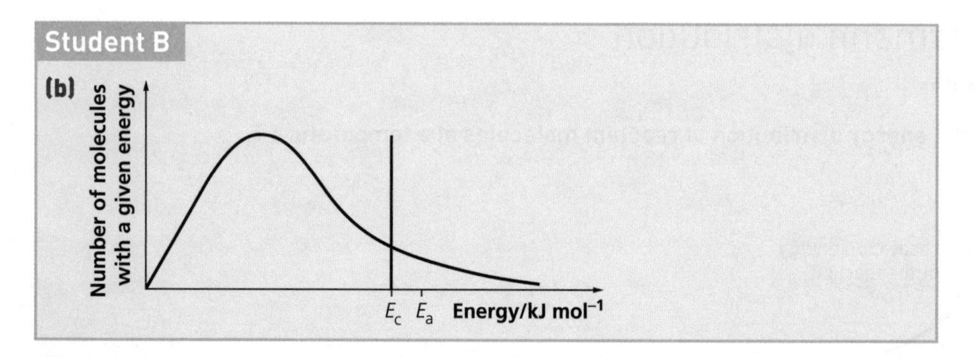

ⓔ Both score the mark. The activation energy for the catalyst must be lower than the original activation energy.

Student A

(c) The mode of action of any catalyst is to lower the activation energy, so that more particles now have enough energy to react.

Student B

(c) A catalyst is a substance that speeds up a reaction without itself being altered. Catalysts can be homogeneous (i.e. the same phase) or heterogeneous (a different phase). Catalysts can be reused.

e Student A scores both marks, but unfortunately Student B gains no marks. Student B has not read the question carefully. He/she has simply written down correct information about catalysts, but nothing relevant to the question.

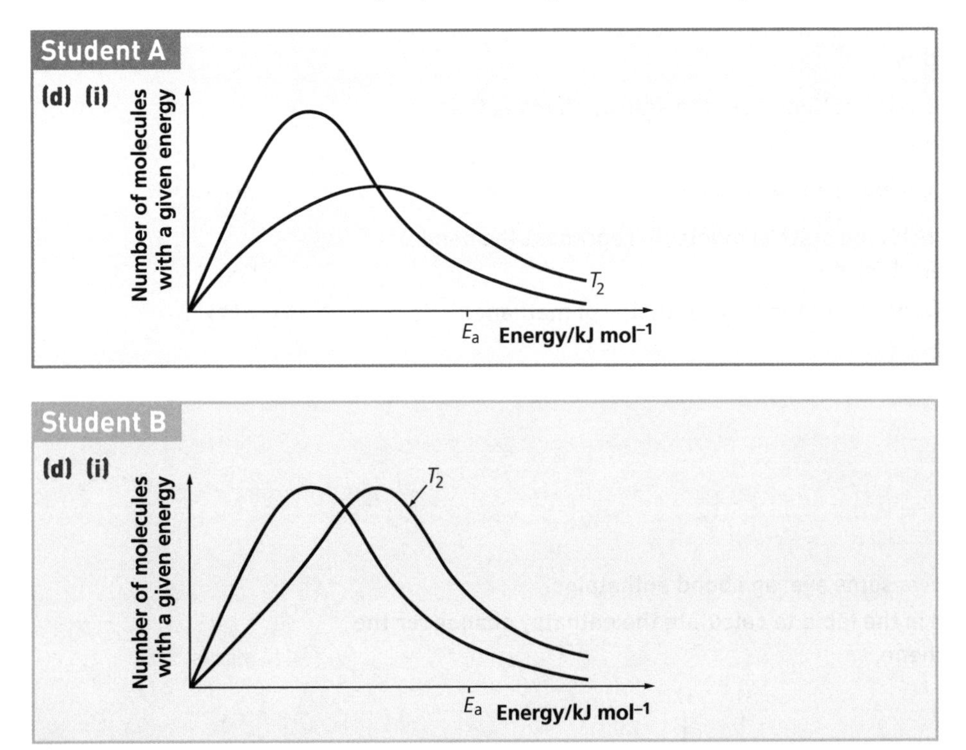

e Student A scores both marks, but Student B drops a mark and only scores 1 out of the 2 possible marks. Both students show that at increased temperature the distribution moves to the right (to higher energy), and hence both score 1 mark. However, as the distribution moves to the right, the curve also flattens out, and hence Student A scores the second mark but Student B doesn't.

Student A

(d) (ii) Increasing temperature increases energy. Therefore more particles will have energy greater than or equal to the minimum energy required, and so the reaction will speed up.

Student B

(d) (ii) Increasing temperature lowers the activation energy, and hence more particles exceed the activation energy.

ⓔ Student A gives the perfect response and gains both marks, but Student B has misunderstood the effect that increasing temperature has on the activation energy, and has probably confused this effect with that of a catalyst. Unfortunately, Student B scores no marks. (Catalysts lower E_a, but changing the temperature has no effect on the value of E_a.)

ⓔ **The outcome is vastly different, with Student A scoring full marks and Student B only scoring 2 out of 8 marks (a grade U).**

Question 15 Bond enthalpy and catalysts

Time allocation: 11–12 minutes

Bond enthalpies can provide information about the energy changes that accompany a chemical reaction.

(a) **What is meant by the term 'bond enthalpy'?** (2 marks)

(b) **(i)** **Write an equation, including state symbols, to represent the bond enthalpy of hydrogen chloride.** (1 mark)

(ii) **Write an equation to represent the bond enthalpy of methane.** (2 marks)

(c)

Bond	Average bond enthalpy/kJ mol^{-1}
C–C	350
C=C	610
H–H	436
C–H	410

(i) **The table above shows some average bond enthalpies.**
Use the information in the table to calculate the enthalpy change for the hydrogenation of ethene. (3 marks)

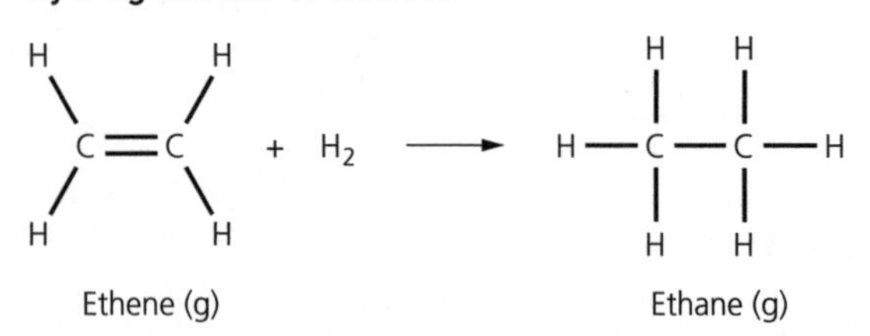

(ii) **The enthalpy change of this reaction can be found by experiment to be −136 kJ mol^{-1}. Explain why this value is different from that determined above.** (2 marks)

(d) **In the above reaction, nickel is used as a catalyst.**

(i) **Define 'catalyst'.** (1 mark)

(ii) **Explain the mode of action of nickel in this reaction.** (3 marks)

Total: 14 marks

e The command terms 'what is meant by' in (a) and 'define' in (d)(i) are in this context equivalent and only a simple statement is required. As in previous cases 'explain' requires reasons to justify your answers.

Student A

(a) It is the enthalpy change when 1 mole of a covalent bond is broken in the gaseous state.

Student B

(a) The energy needed to break 1 mole of a bond in the gas state.

e This term is difficult to define, but there are certain key elements that must be there, namely 'breaking a bond', and it must be in the 'gaseous state', and must involve '1 mole'. It is probably helpful to illustrate the definition with the equation: $X{-}Y(g) \rightarrow X(g) + Y(g)$, which shows that you also know that it relates to covalent molecules and involves homolytic fission. However, both students score both of the marks.

Student A

(b) (i) $HCl(g) \rightarrow H(g) + Cl(g)$

Student B

(b) (i) $HCl \rightarrow H + Cl$

e Student A scores the mark, but Student B loses the mark, because the state symbols are essential.

Student A

(b) (ii) $CH_4(g) \rightarrow CH_3(g) + H(g)$

Student B

(b) (ii) $CH_4(g) \rightarrow C(g) + 4H(g)$

e Neither student scores both marks. Student A has taken the definition of bond enthalpy literally and has 'broken 1 bond in the gaseous state', but because methane has four C–H bonds, the bond enthalpy is the average of all four bond enthalpies. Student A would be awarded 1 mark. Student B has shown the enthalpy change for breaking four C–H bonds, not one, and would also only score 1 mark.

The required equation is: ¼[$CH_4(g) \rightarrow C(g) + 4H(g)$], which clearly indicates that the bond enthalpy in methane is the average bond enthalpy taken when all four C–H bonds are broken. This is shown below:

$CH_4(g) \rightarrow CH_3(g) + H(g)$ $\Delta H = +425\,kJ\,mol^{-1}$

$CH_3(g) \rightarrow CH_2(g) + H(g)$ $\Delta H = +470\,kJ\,mol^{-1}$

$CH_2(g) \rightarrow CH(g) + H(g)$ $\Delta H = +416\,kJ\,mol^{-1}$

$CH(g) \rightarrow C(g) + H(g)$ $\Delta H = +335\,kJ\,mol^{-1}$

The total enthalpy change is (+425 + 470 + 416 + 335) = +1646 $kJ\,mol^{-1}$, which is the enthalpy change when four C–H bonds are broken in methane. Therefore, the enthalpy change when one C–H bond is broken is +1646/4 = 411.5 $kJ\,mol^{-1}$. The students clearly are not expected to quote numerical values, but they are expected to realise that the bond enthalpy quoted is the average value.

Student A

(c) (i) Bonds broken (610 + 1640 + 436) = +2686

Bonds formed (−350 − 2460) = −2810

Enthalpy change = −124 $kJ\,mol^{-1}$

Student B

(c) (i) Bonds broken: 1 × C=C + 610

Bonds formed: 2 × C–H −820

1 × C–C −350

Enthalpy change: −560 $kJ\,mol^{-1}$

e There are two ways of carrying out this calculation. Student A has opted for the safer way and calculated the enthalpy change for breaking every bond in the reactants and then for forming every bond in the reactant. A sensible way to set this out is to simply list each and every bond:

Bonds broken:

1 × C=C + 610

4 × C–H +1640

1 × H–H +436

Bonds formed:

1 × C–C −350

6 × C–H −2460

Enthalpy change = −124 $kJ\,mol^{-1}$

Student B has looked at the overall change and has attempted to identify the net change in bonds broken and formed. Student B has correctly worked out that if four C–H bonds are broken and six C–H bonds are formed, then the net change is the formation of two C–H bonds. However, Student B has forgotten that H_2 consists of an H–H bond that also has to be broken. Student A scores all 3 marks, and Student B gains 2 marks even though the answer is incorrect. This will be marked 'consequentially', as there is only one error in the calculation, and therefore Student B will lose only 1 mark.

Student A

(c) (ii) The bond energies used in the calculation are average values for the bonds.

Student B

(c) (ii) Experiments are not very accurate and heat will be lost.

e Student A gains 1 mark and would have scored the second had he/she gone on to explain that the C–H bond in ethene will not be the same as the C–H bond in ethane, because they are in different environments. Student B has made unjustified assumptions about the accuracy of the experiment and scores no marks.

Student A

(d) (i) Speeds up a reaction without being used up.

Student B

(d) (i) Speeds up a reaction by lowering the activation energy.

e Both score the mark.

Student A

(d) (ii) The ethene and hydrogen gases are absorbed by the Ni, the reaction takes place, and the ethane is desorbed.

Student B

(d) (ii) The reactants bind to the surface of the Ni (adsorb) and the bonds are weakened. The reaction takes place and the product leaves the surface of the Ni (desorbs).

ⓔ The marking points are:

- adsorbs to Ni surface ✓
- weakens bonds/lowers activation energy ✓
- desorbs from Ni surface ✓

Student A only scores 1 out of 3 marks. By using the word 'absorbed', the first marking point is lost. In addition, no reference is made to how the bonds are weakened. Student B gains all 3 marks.

ⓔ **Student A scores 10 out of 14 marks, which is a B grade (71%). Student B scores only 1 mark less, but this makes it a C-grade answer (64%).**

Question 16 Enthalpy of combustion, Hess's law, catalytic converters

Time allocation: 9–10 minutes

Octane, C_8H_{18}, is one of the hydrocarbons present in petrol.

(a) **Define the term 'standard enthalpy change of combustion'.** (3 marks)

(b) **Use the data in the following table to calculate the standard enthalpy change of combustion of octane:**

$$C_8H_{18}(l) + 12\tfrac{1}{2}O_2(g) \rightarrow 8CO_2(g) + 9H_2O(l)$$

Compound	$\Delta_f H^{\ominus}$/kJ mol^{-1}
$C_8H_{18}(l)$	−250.0
$CO_2(g)$	−393.5
$H_2O(l)$	−285.9

(3 marks)

(c) **Combustion in a car engine also produces polluting gases, mainly carbon monoxide, unburnt hydrocarbons and oxides of nitrogen such as nitrogen(II) oxide, NO. Explain, with the aid of equations, how CO and NO are produced in a car engine.** (2 marks)

(d) **(i)** **The catalytic converter removes much of this pollution in a series of reactions. Write an equation showing the removal of carbon monoxide and nitrogen monoxide gases.** (1 mark)

(ii) **The removal of carbon monoxide and nitrogen monoxide gases involves a redox reaction. Use your answer to (d)(i) to identify the element being reduced and state the change in its oxidation number.** (2 marks)

Total: 11 marks

ⓔ The command word 'define' in (a) requires only a simple statement. In (c), 'explain' requires only balanced equations as directed. In (d)(ii) 'identify' allows some flexibility: the name or formula is OK, but the identification must be unambiguous.

Student A

(a) It is the enthalpy change when 1 mole of a substance is burnt completely, in an excess of oxygen, under standard conditions of 298 K and 1 atmosphere.

Student B

(a) It's the enthalpy change when 1 mole of a substance is burnt in oxygen, under standard conditions.

ⓔ Student A gets all 3 marks. The marking points are:

- 1 mole ✓
- burnt in an excess of oxygen ✓
- standard conditions are 298 K/25°C and 100 kPa/1 atm ✓

Student B only scores 1 mark. Failing to specify the standard conditions is careless and demonstrates poor examination technique rather than a lack of knowledge. It is essential that excess oxygen is used to avoid incomplete combustion, which can produce carbon monoxide.

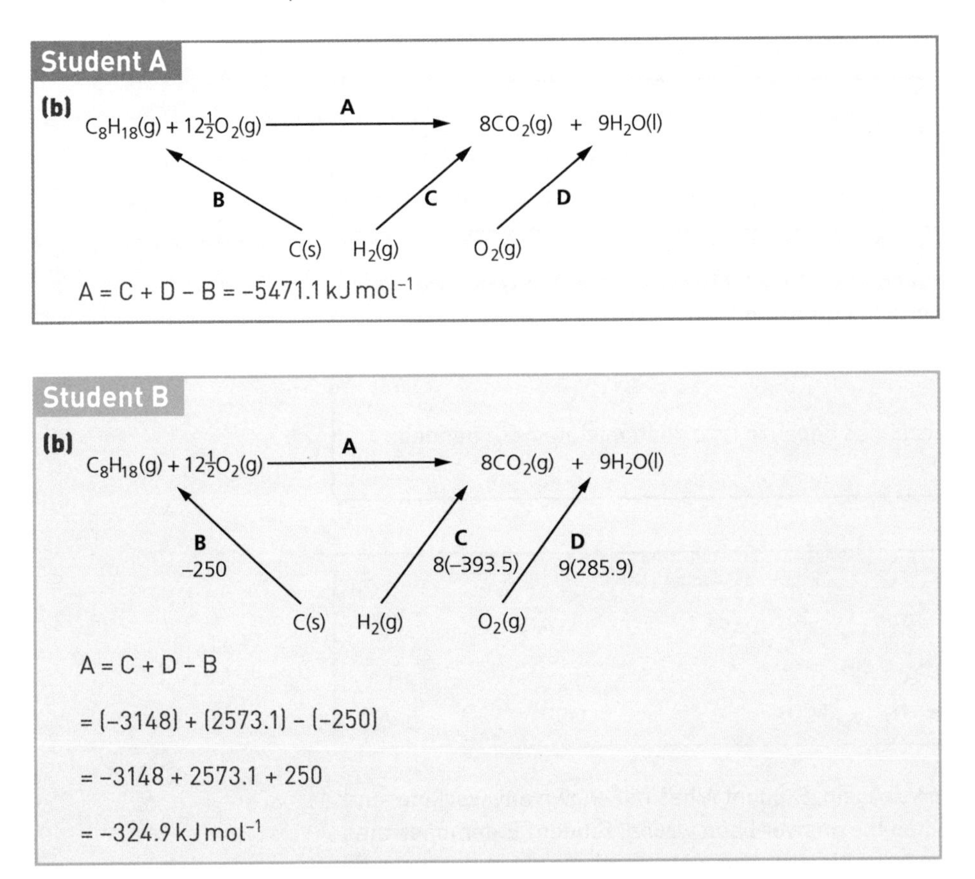

ⓔ Student B shows better examination technique by giving all the working, but has forgotten to write the minus sign in front of the 285.9 for the formation of water. This error is clear and can be tracked through the working. Student B would therefore gain 2 of the 3 marks available. Although Student A scores all 3 marks, had the numerical value been incorrect, he/she would have only scored 1 mark or possibly zero. *Remember:* it is always better to show all of your working in any calculation.

Student A

(c) $C_8H_{18}(l) + 8\frac{1}{2}O_2(g) \rightarrow 8CO(g) + 9H_2O(l)$

$N_2(g) + O_2(g) \rightarrow 2NO(g)$

Student B

(c) $C_8H_{18} + 9\frac{1}{2}O_2 + N_2 \rightarrow 8CO + 9H_2O + 2NO$

e Both students score 2 marks. Student B's response is a little unusual, but nevertheless correct.

Student A

(d) (i) $2NO(g) + 2CO(g) \rightarrow N_2(g) + 2CO_2(g)$

Student B

(d) (i) $2NO + 2CO \rightarrow N_2 + 2CO_2$

e Both students gain the mark. Marks for state symbols will only be awarded/deducted if they are asked for in the question.

Student A

(d) (ii) Nitrogen has been reduced because its oxidation state has changed from +2 to zero.

Student B

(d) (ii)

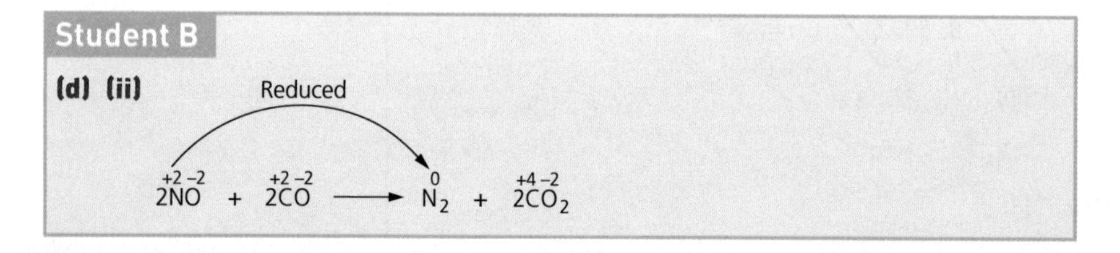

e Both students gain 2 marks. Again, Student A has not shown any working, and would have lost both marks had the answer been wrong. Student B demonstrates better examination technique by writing the oxidation numbers along the top of the equation, so that the examiner can follow the working.

e **Both students have done well, and Student A has scored full marks. However, Student B carelessly lost 3 marks and scored 8 out of 11 marks (71%), which is a grade B.**

Question 17 Equilibria

Time allocation: 12–13 minutes

Sulfuric acid, H_2SO_4, is made industrially by the Contact process. This reaction is an example of a dynamic equilibrium:

$$2SO_2(g) + O_2(g) \rightleftharpoons 2SO_3(g) \qquad \Delta H = -98\,kJ\,mol^{-1}$$

(a) State two features of a reaction with a *dynamic equilibrium*. (2 marks)

(b) State and explain what happens to the equilibrium position of this reaction as:

(i) the temperature is raised (2 marks)

(ii) the pressure is increased (2 marks)

(iii) Suggest the optimum conditions for the Contact process. (2 marks)

(c) **(i)** The conditions used for the Contact process are a temperature of 450°C to 600°C and a pressure of around 10 atmospheres.

Explain why the optimum conditions are not used. (3 marks)

(ii) Vanadium(v) oxide is used as a catalyst. What effect does this have on the conversion of $SO_2(g)$ into $SO_3(g)$? (2 marks)

(iii) At least three catalyst chambers are used to ensure maximum conversion of $SO_2(g)$. The conversion yield can exceed 98%.

State two advantages of this high conversion rate. (2 marks)

Total: 15 marks

e The command word 'suggest' in (b)(iii) indicates that you are not expected to recall the values but you are meant to predict them using your answers from (b)(i) and (b)(ii).

Student A

(a) The rates of both reactions are the same, and therefore the amount of each chemical remains the same.

Student B

(a) The amount of each chemical in the system remains constant.

The reagents react at the same rate as the products.

e Both students gain both marks.

Student A

(b) (i) The position of the equilibrium moves to the left because the forward reaction is exothermic.

Student B

(b) (i) The reaction is much faster because more particles now exceed the activation energy.

e Student A scores both marks and has sensibly used the same words — 'equilibrium position' — as those used in the question. Unfortunately, Student B has either misread or misunderstood the question, and has explained the effect of increasing temperature on the rate of reaction. Student B's explanation is correct, but it scores no marks because it does not address the question that was set. This is a common error, and many students lose valuable marks by not reading the question carefully.

Student A

(b) (ii) The equilibrium moves to the right because there are fewer molecules of gas on the right-hand side.

Student B

(b) (ii) Increasing pressure effectively increases the concentration, and therefore the reaction will go faster.

e Student B has again misread the question and made the same mistake as in (b)(i). This has cost another 2 marks.

Student A

(b) (iii) Low temperature and high pressure

Student B

(b) (iii) Low temperature and high pressure

e Both students gain 2 marks. This shows that Student B clearly understands equilibrium and should have gained all 4 marks for (b)(i) and (b)(ii).

Student A

(c) (i) At low temperature the conversion is high, but the rate of reaction is too slow. High pressure is too expensive.

Student B

(c) (i) Temperature: a compromise is reached between rate and conversion. At low temperature the rate is too slow.
Pressure: a compromise is reached between cost and conversion.
Catalyst: a catalyst is used to speed up the rate of conversion, so that it is cost effective to work at a low pressure.

ⓔ Student B gives the perfect answer and is awarded all 3 marks. Student B's response to this section demonstrates understanding and indicates that he/she should not have lost the marks in part (b). Student A just about scores 2 marks. The explanation of temperature is fine, but the explanation regarding pressure is barely adequate. As well as the use of a catalyst, the key to the answer is the compromise between the rate and percentage yield of SO_3 and the cost and percentage yield of SO_3.

Student A

(c) (ii) The catalyst speeds up the reaction, but it doesn't change the equilibrium position because it speeds up the forward and the reverse reactions equally.

Student B

(c) (ii) It speeds up the reaction by providing a mechanism of lower activation energy.

ⓔ Student A scores 2 marks, but again Student B seems to have rushed the answer and has not fully addressed the question, scoring only 1 mark.

Student A

(c) (iii) More cost efficient and reduces the amount of SO_2 pollution.

Student B

(c) (iii) More profitable and can make more money.

ⓔ Student A gives a very good answer and gains both marks, but Student B scores only 1 mark, having written the same thing twice.

ⓔ **Student B scores 9 marks for this question, which is just a grade C. However, if the student had not lost 4 marks in (b)(i) and (ii), then a grade A would have been awarded. Every mark is important, and it is essential to read the question carefully. Again, Student A gives a good answer, scoring 14 out of 15 marks.**

Question 18 Activation energy, le Chatelier's principle

Time allocation: 13–15 minutes

Hydrogen can be reacted with nitrogen in the presence of a catalyst to make ammonia, using the Haber process.

$$N_2(g) + 3H_2(g) \rightleftharpoons 2NH_3(g)$$

The activation energy for the forward reaction is +68 kJ mol^{-1}.

The activation energy for the reverse reaction is +160 kJ mol^{-1}.

(a) (i) Use this information to sketch the energy-profile diagram. Label clearly the activation energy for the forward reaction, E_f, and the activation energy for the reverse reaction, E_r. (2 marks)

(ii) Explain what is meant by 'activation energy'. (1 mark)

(iii) Calculate the enthalpy change for the forward reaction. (1 mark)

Much of the ammonia produced is oxidised into nitric acid, using the Ostwald process, which involves three stages:

Stage 1 $4NH_3(g) + 5O_2(g) \rightleftharpoons 4NO(g) + 6H_2O(g)$ $\Delta H = -950\,kJ\,mol^{-1}$

Stage 2 $2NO(g) + O_2(g) \rightleftharpoons 2NO_2(g)$ $\Delta H = -114\,kJ\,mol^{-1}$

Stage 3 $3NO_2(g) + H_2O(g) \rightleftharpoons 2HNO_3(g) + NO(g)$ $\Delta H = -117\,kJ\,mol^{-1}$

(b) (i) With reference to the oxidation number of the nitrogen in $NH_3(g)$ (stage 1) and HNO_3 (stage 3), show clearly that this is an oxidation process. (3 marks)

(ii) State le Chatelier's principle. (2 marks)

(iii) In stage 1, use le Chatelier's principle to predict and explain the temperature and the pressure that would give the maximum yield at equilibrium. (4 marks)

(iv) Suggest what happens to the NO(g) produced in stage 3. (1 mark)

(c) The nitric acid produced in stage 3 is a strong acid. Explain, with the aid of an equation, what is meant by the term strong acid. (2 marks)

Total: 16 marks

e The command word 'sketch' in (a)(i) clearly requires a diagram, but it is essential to ensure that the diagram is labelled, including any units.

Student A

(a) (i)

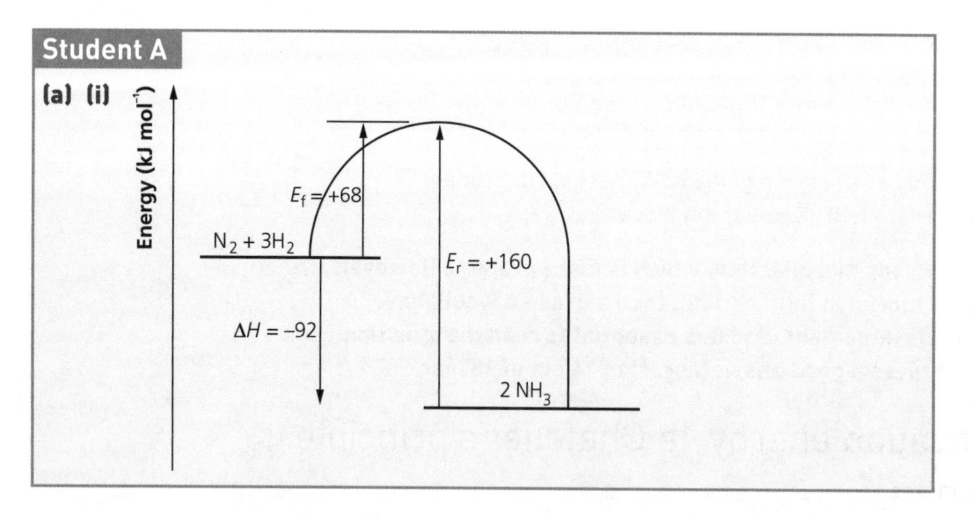

Student B

(a) (i)

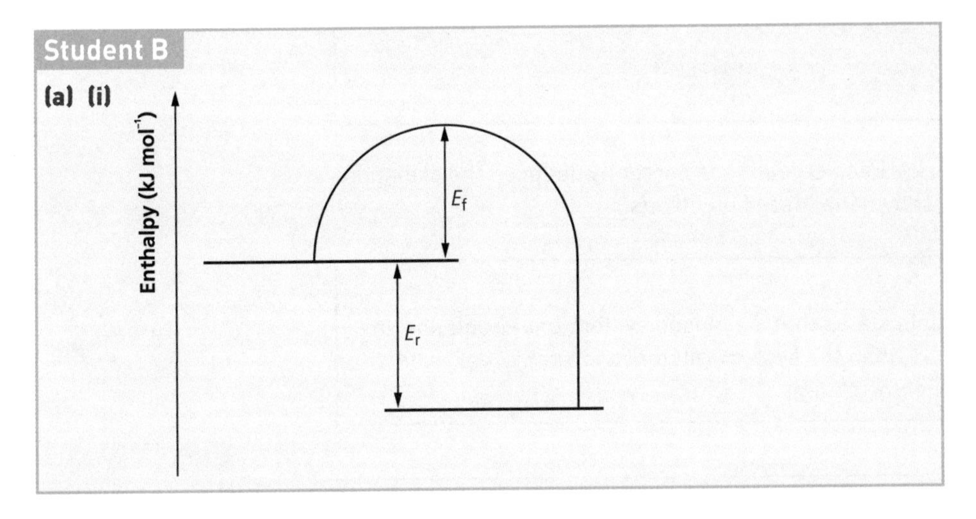

ⓔ The marking points are:

- products at a lower enthalpy than the reactants ✓
- activation energy of both forward and reverse reactions correctly labelled ✓

Student A gives the perfect answer and even goes on to work out ΔH for (a)(iii). Student B scores no marks. The activation energy for the reverse reaction is incorrect. Moreover, the energy levels are not labelled, so it is not possible to award the first marking point.

Student A

(a) (ii) Activation energy is the minimum energy required to start a reaction.

Student B

(a) (ii) It's the minimum energy needed to start the reaction.

ⓔ Both students gain the mark. The key phrase is the *minimum* energy.

Student A

(a) (iii) $-92\,kJ\,mol^{-1}$

Student B

(a) (iii) 92

ⓔ Student A gains the mark, but Student B doesn't score anything because the sign of the enthalpy change for the forward reaction is negative.

Student A

(b) (i) The N in NH_3 is −3 and the N in HNO_3 is +5. An increase in oxidation number involves a loss of electrons, which is oxidation (OILRIG).

Student B

(b) (i) −3 to +5 $N^{-3} \rightarrow N^{+5} + 8e^-$

Ⓔ Both students gain 3 marks. Each student has correctly deduced the oxidation numbers, and has stated or shown the loss of electrons.

Student A

(b) (ii) le Chatelier's principle states that if a closed system under equilibrium is subject to a change, then the system will move in such a way as to minimise the effect of the change.

Student B

(b) (ii) If concentration, temperature and pressure are changed, the system will move to oppose the change. If we increase the concentration, the system will move in such a way as to decrease the concentration. If we increase the temperature, the system will move in such a way as to decrease the temperature. If we decrease the pressure, the system will try to move in such a way as to increase the pressure.

Ⓔ Both students gain 2 marks. Student A gives a textbook answer, while Student B fully explains le Chatelier's principle.

Student A

(b) (iii) The forward reaction is exothermic and so will be favoured by low temperature. There are fewer products than reactants, so high pressure is required.

Student B

(b) (iii) Low temperature and high pressure, but in industry a temperature of about 450°C and a pressure of 200 atm are used.

Ⓔ Student A scores 2 out of 4 marks. The prediction and explanation about temperature are correct and earn 2 marks. Student A concludes incorrectly that there are fewer products than reactants (the reverse is true), thus losing both marks relating to the required pressure. Student B gains 2 marks for the correct conditions, but doesn't score any marks for explaining why a low temperature is needed. Student B has simply memorised the industrial conditions.

Student A

(b) (iv) Reused in stage 2

Student B

(b) (iv) Recycled

Both students gain the mark.

Student A

(c) Strong acids dissociate totally, e.g. $HCl \rightarrow H^+ + Cl^-$

Student B

(c) All acids have a pH below 7. Strong acids have a low pH of about 1 or 2.

This question is worth 2 easy marks, but neither student scores both marks. You need to explain both 'strong' and 'acid'. Student A states that strong acids dissociate totally, for 1 mark, but does not explain that acids are proton donors. If the student had written $HCl + H_2O \rightarrow H_3O^+ + Cl^-$, he/she would have scored both marks. Student B misses the point and gives a property of acids without addressing the question. He/she would probably score no marks.

Student A scores 12 out of 16 marks (75% — a grade B answer), while Student B gains 9 out of 16 (56%, a grade D).

Question 19 Alkanes and haloalkanes

Time allocation: 6–7 minutes

(a) Ethane, C_2H_6, reacts with chlorine (Cl_2) in the presence of sunlight to form a mixture of chlorinated products. One possible product is $C_2H_4Cl_2$.

(i) State the type of mechanism involved in this reaction. (1 mark)

(ii) The initiation step involves the homolytic fission of the Cl–Cl bond. What is meant by the term homolytic fission? (1 mark)

(b) Name the two possible isomers of $C_2H_4Cl_2$. (2 marks)

(c) When $C_2H_4Cl_2$ is treated with aqueous NaOH, it undergoes substitution reactions to form both $C_2H_4(OH)Cl$ and $C_2H_4(OH)_2$.

(i) State, and explain, the role of the $OH^-(aq)$ in these reactions. (2 marks)

(ii) Draw two possible isomers of $C_2H_4(OH)_2$. (2 marks)

Total: 8 marks

The command word 'state' used in part (a)(i) indicates that a brief answer is required with no supporting argument. In (c) 'state and explain' requires a simple standard definition and also an explanation of the initial statement. In (b) 'name' means what it says. Oddly a substantial number of students ignore the instructions and draw the structures of the two isomers.

Student A

(a) (i) Radical substitution

Student B

(a) (i) Radical

ⓔ Student A gets 1 mark, but Student B scores no marks. The key phrase is 'radical substitution'.

Student A

(a) (ii) The bond breaks (i.e. fission), so that each atom in the bond retains one of the bonded electrons. For example:

$Cl–Cl \rightarrow 2Cl\bullet$

Student B

(a) (ii) Radicals are produced by homolytic fission.

ⓔ Student A gives a perfect answer and scores 1 mark, while Student B scores no marks. Student B has missed the point of the question and has stated a consequence of homolytic fission rather than explaining what is *meant* by homolytic fission.

Student A

(b) 1,2-dichloroethane and 1,1-dichloroethane

Student B

(b) 1,1-dichloroethane and 1,2-dichloroethane

ⓔ Both answers are correct, for 2 marks.

Student A

(c) (i) Nucleophile, because it donates an electron pair to the $C^{\delta+}$.

Student B

(c) (i) OH^- is a nucleophile.

ⓔ Student A gets 2 marks, but Student B only scores 1 mark for *stating* that it is a nucleophile. The clue is in the question, and a simple *explanation* is required for the second mark.

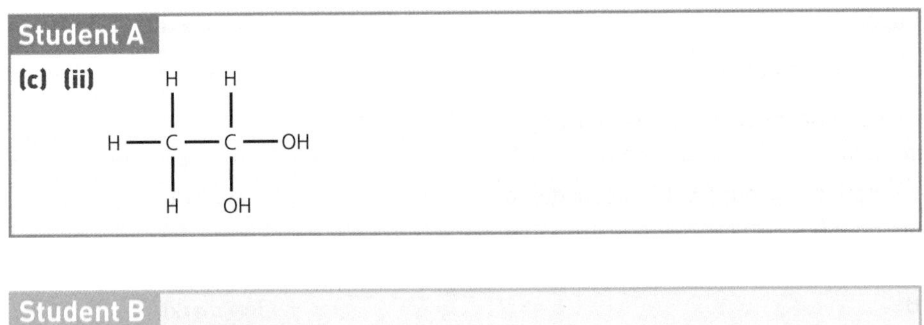

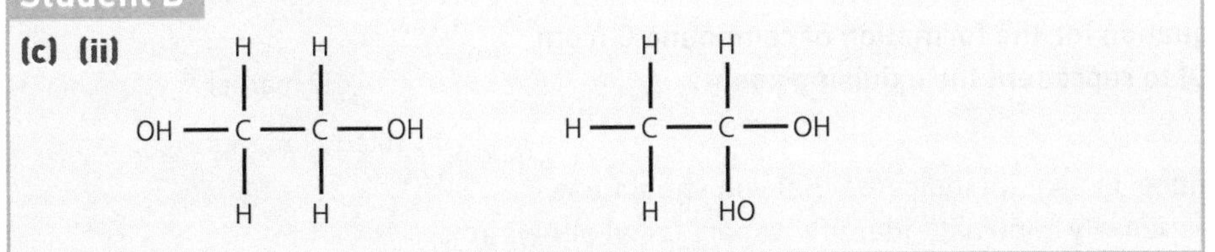

ⓔ Student A scores only 1 mark, because he/she has not read the instructions carefully and has drawn only one isomer. Student B scores 1 mark, but could have lost both marks. Student B recognises that it is possible to form ethane-1,2-diol and ethane-1,1-diol (although the latter doesn't exist), but then loses a mark by carelessly drawing the C–OH bond to the H and not to the O, i.e. C–HO. This is known as a bond-linkage error and is usually only penalised once in each paper.

The isomers should have been drawn as:

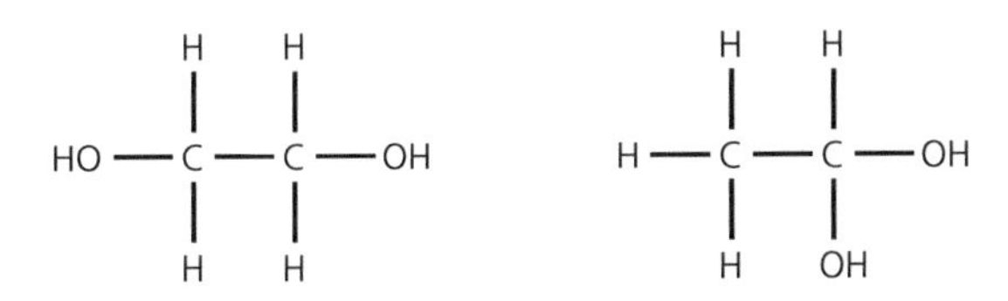

ⓔ **Student A does well and scores 7 marks out of 8, but carelessly loses a mark in the final part. Student B only gains 4 out of 8 marks. Student B's overall response places him or her as borderline grade D, but with a little care this could have been pushed up by 2 or 3 marks, to the A/B borderline.**

Question 20 Alcohols

Time allocation: 9–10 minutes

Compound A contains two functional groups.

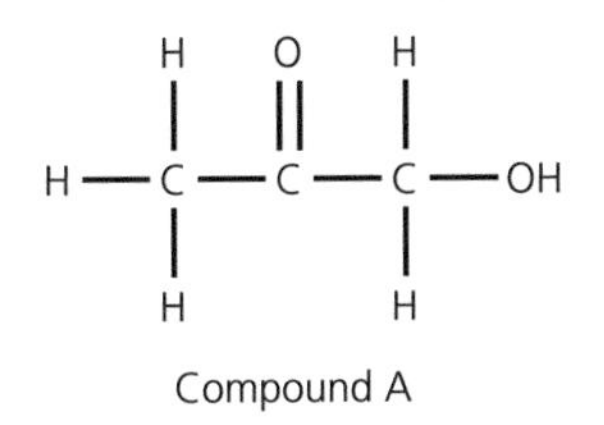

Compound A

(a) Name the functional groups. (3 marks)

(b) Deduce the molecular formula of compound A. (1 mark)

(c) Compound A can be oxidised to produce a mixture of compound B, molecular formula $C_3H_4O_2$, and compound C, molecular formula $C_3H_4O_3$.

(i) Identify which of the functional groups could be oxidised. (1 mark)

(ii) State a suitable oxidising mixture. (2 marks)

(iii) State what you would observe during the oxidation. (1 mark)

(iv) Identify compound B. (1 mark)

(v) Write a balanced equation for the formation of compound C from compound A. Use [O] to represent the oxidising agent. (2 marks)

Total: 11 marks

ⓔ The command word 'deduce' in part (b) indicates that you should use information given in the question. By contrast, 'identify' in part (c)(iv) allows you flexibility — you could either name compound B or identify it by its structural, displayed or skeletal formula. The identification must be unambiguous such that the molecular formula would not score the mark.

Student A

(a) Primary alcohol and a ketone

Student B

(a) Ketone and an alcohol

ⓔ Student A scores all 3 marks but Student B drops 1 mark by not *classifying* the alcohol. It is important that you use the information in the question. Each part of a question will tell you how many marks are allocated to that part, and it is essential that you use this information. Here there are 3 marks available, so three points are needed: ketone ✓, primary ✓ and alcohol ✓.

Student A

(b) $C_3H_6O_2$

Student B

(b) CH_3COCH_2OH

ⓔ Student A gets the mark, but Student B has given a *structural* formula rather than a *molecular* formula. Molecular formulae always group together all atoms of the same element, and they are always written in the form: $C_xH_yO_z$. Student B gains no marks.

Student A

(c) (i) Alcohol

Student B

(c) (i) Alcohol

ⓔ Each student gains the mark.

Student A

(c) (ii) Acidified dicromate

Student B

(c) (ii) $H^+/Cr_2O_7^-$

ⓔ Student A gets both marks, while student B loses a mark by using an incorrect formula for the dichromate ion. Student A spelt dichromate incorrectly, but it is highly unlikely that this would be penalised. However, the same tolerance is not extended to incorrect formulae. The formula should be $Cr_2O_7^{2-}$; Student B didn't include the correct charge on the ion and therefore loses the mark.

Student A

(c) (iii) Colour change

Student B

(c) (iii) Turns green

ⓔ Neither student scores the mark. Student A's response is too vague. Student B hasn't stated the full colour change from orange to green.

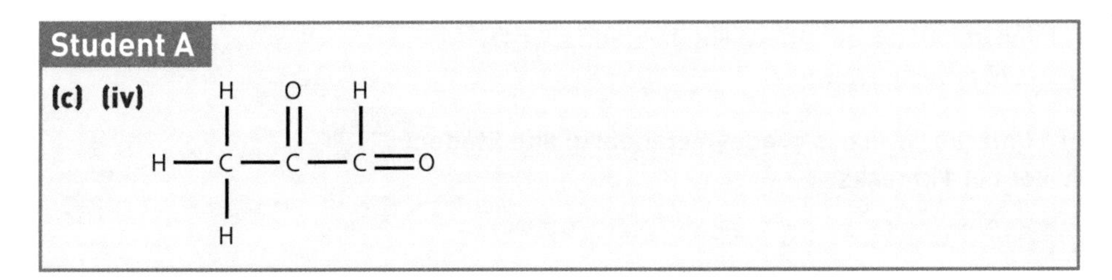

Student B

(c) (iv) B is CH_3COCOH

ⓔ Student A gains the mark, but Student B doesn't score. Student A has made good use of the information in the question and used the displayed formula of compound A to deduce the structure of B. Student B's response is also good, but perhaps a little too clever. The final part of CH_3COCOH indicates an alcohol. If it had been written as CH_3COCHO, it would have scored the mark. Student B clearly has potential but needs to take more care.

Student A

(c) (v)

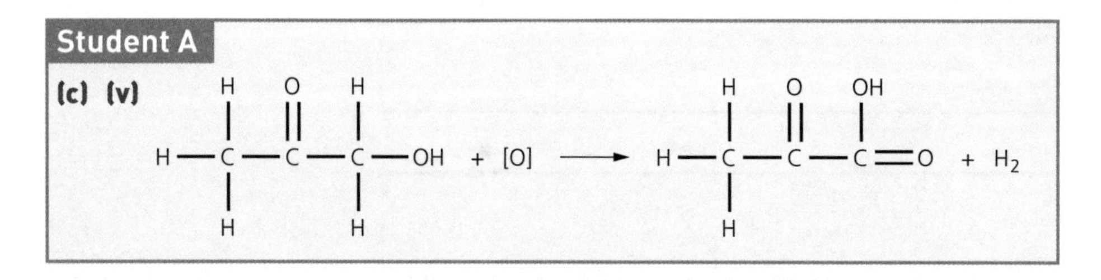

Student B

(c) (v) $C_3H_6O_2 + 2[O] \rightarrow C_3H_4O_3 + H_2O$

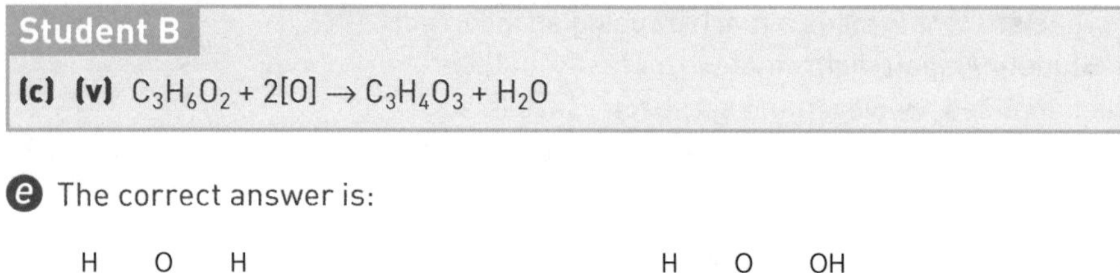

ⓔ The correct answer is:

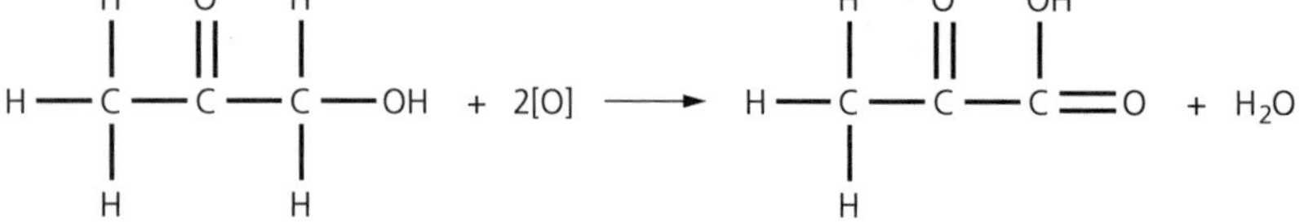

The first mark would be awarded for the correct organic product and the second mark for balancing the equation.

Both students would score 1 out of 2 marks. Student A is very careful and makes good use of information in the question, but H_2 is never formed in the presence of an oxidising agent; water will always be produced. Student B is trying to be too clever. The equation is essentially correct and balanced, but, by using molecular formulae, the product hasn't been correctly identified. The formula $C_3H_4O_3$ is ambiguous and could be any one of several isomers.

ⓔ **Student A scores 9 out of 11 marks (which is grade-A standard) and Student B would earn a grade E with 5 out of 11 marks.**

Question 21 Radical and nucleophilic substitution

Time allocation: 15–16 minutes

Ethane, C_2H_6, reacts with Cl_2 in the presence of sunlight to form a mixture of chlorinated products. One possible product is C_2H_5Cl, formed as shown in the following equation:

$$C_2H_6 + Cl_2 \rightarrow C_2H_5Cl + HCl$$

(a) Describe, with the aid of equations, the mechanism of this reaction. (4 marks)

(b) One other possible product of the reaction between ethane and chlorine is compound A.

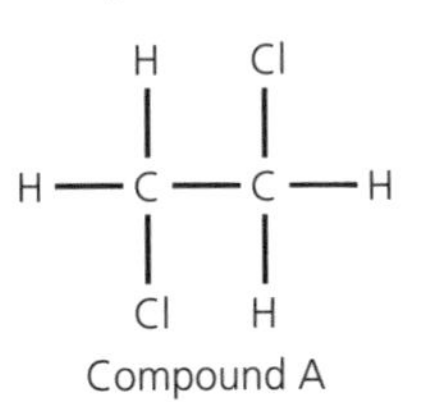

Compound A

(i) Name compound A. (1 mark)

(ii) Draw an isomer of compound A. (1 mark)

(c) Chloroethane can react with a solution of sodium hydroxide, as follows:

$$C_2H_5Cl + OH^- \rightarrow C_2H_5OH + Cl^-$$

(i) State the solvent in which the sodium hydroxide is dissolved. (1 mark)

(ii) State and explain the role of the hydroxide ion (OH^-) in this reaction. (2 marks)

(d) Ethanol, C_2H_5OH, is refluxed with an acidified solution of potassium dichromate(VI) to produce ethanoic acid. The acidified potassium dichromate(VI) acts as an oxidising agent.

(i) Explain what is meant by the term *reflux*. (1 mark)

(ii) State what colour changes take place in the reaction mixture. (1 mark)

(iii) Write a balanced equation for the oxidation of ethanol to ethanoic acid. The oxidising agent can be represented as [O] in your equation. (2 marks)

(e) An infrared spectrum of propanoic acid was obtained (see below). By referring to your datasheet, identify two peaks in the infrared spectrum that confirm the presence of the carboxylic acid functional group. (4 marks)

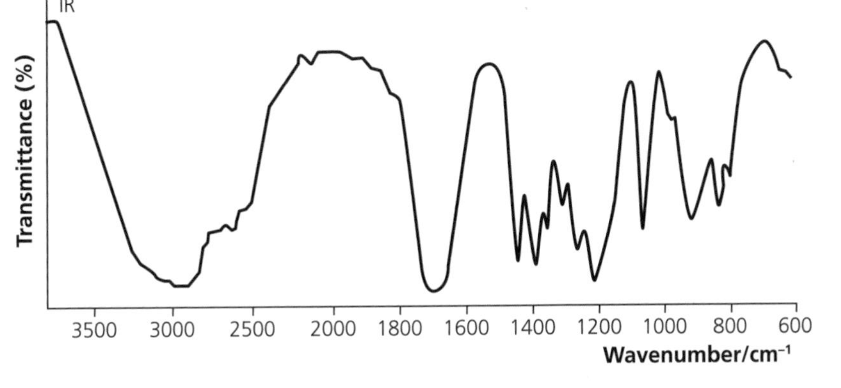

Total: 17 marks

e The command word 'describe' requires significant detail. When referring to a mechanism, equations are required and the movement of electrons must be tracked. In the radical substitution mechanism this is done by using a dot to track the formation and reaction of the radicals. In (e) make sure you follow the instructions and use your datasheet.

Student A

(a) $Cl_2 \rightarrow 2Cl\bullet$

$C_2H_6 + Cl\bullet \rightarrow HCl + C_2H_5\bullet$

$C_2H_5\bullet + Cl_2 \rightarrow C_2H_5Cl + Cl\bullet$

$Cl\bullet + Cl\bullet \rightarrow Cl_2$

Student B

(a) $Cl_2 \rightarrow 2Cl\bullet$

$C_2H_6 + Cl\bullet \rightarrow H\bullet + C_2H_5Cl$

$H\bullet + Cl_2 \rightarrow HCl + Cl\bullet$

$C_2H_5\bullet + Cl\bullet \rightarrow C_2H_5Cl$

e Student A gains all 4 marks, while Student B only picks up the first and last marks. The propagation steps shown by Student B are incorrect. The initial propagation step of the Cl• with any alkane *always* produces an alkyl radical and HCl.

Student A

(b) (i) 1,2-dichloroethane

Student B

(b) (i) 1,2-dichloroethane

e Correct, for 1 mark.

Student A

(b) (ii)

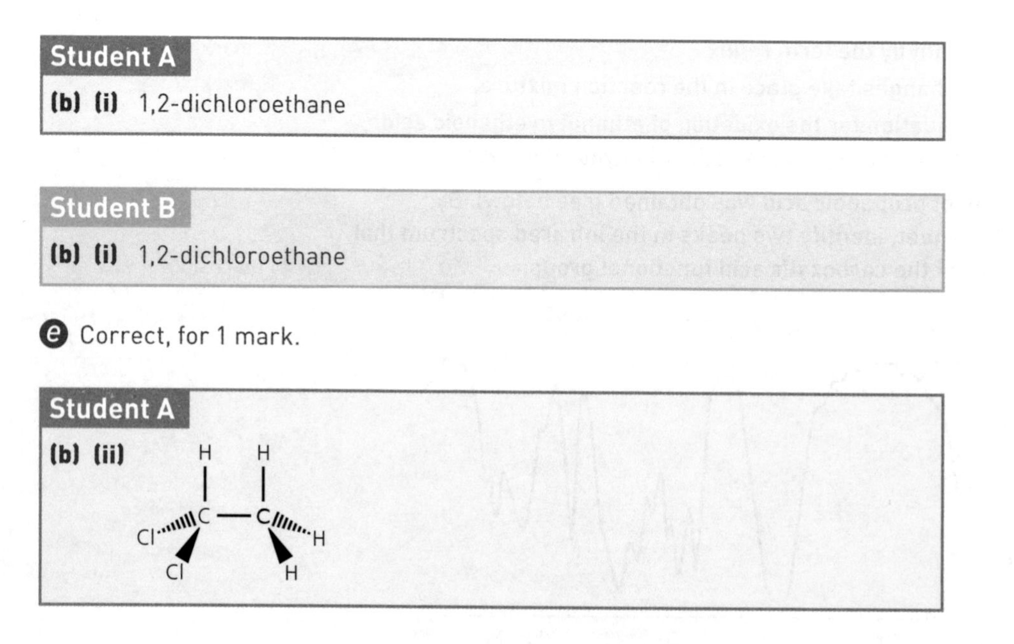

Student B

(b) (ii)

```
    Cl  Cl
    |   |
H — C — C — H
    |   |
    H   H
```

Z-isomer (*cis*)

ⓔ Student A gains the mark, but Student B has fallen into a trap. It is not possible to get *E*/*Z* isomerism (*cis–trans*) unless there is a C=C double bond. The isomer drawn by Student B is identical to the one given in the question, because the atoms can rotate about the C–C single bond.

Student A

(c) (i) Water

Student B

(c) (i) (aq)

ⓔ Both answers score the mark.

Student A

(c) (ii) A nucleophile

Student B

(c) (ii) Donates a lone pair of electrons

ⓔ Each answer is worth 1 out of 2 marks. Neither student has used the information in the question. There are 2 marks available and therefore two separate points are required: *state* that it behaves as a nucleophile ✓ and *explain* that it is because it donates a lone pair of electrons ✓.

Student A

(d) (i) Heat with a condenser

Student B

(d) (i) Evaporates and then condenses

ⓔ Neither student scores the mark. To get the mark you must state that refluxing involves *continuous* evaporation and condensation, so that the volatile reagents/products are retained in the reaction vessel. Student B is almost correct, but has missed the key word *continuous*. Student A recalls that a condenser is needed, but is too vague to gain a mark. If the description is difficult to put into words, then you might want to draw a diagram instead. This would also score the marks.

Student A

(d) (ii) Orange to green

Student B

(d) (ii) Turns green

ⓔ Student A gains the mark, while Student B again loses the mark by only giving the final colour. When recording observations, it is important to write down the initial observation as well as the final observation.

Student A

(d) (iii) $C_2H_5OH + [O] \rightarrow CH_3CO_2H + H_2O$

Student B

(d) (iii) $C_2H_6O + [O] \rightarrow C_2H_4O_2 + H_2$

ⓔ Student A gains 1 mark for the correct products, but loses a mark because the equation isn't balanced. Student B loses both marks and writes the most common incorrect answer. The formula $C_2H_4O_2$ is ambiguous and doesn't clearly identify the organic product as ethanoic acid. At first glance, it looks good because it is balanced, *but* H_2 is not a product. Water is *always* formed. The correct equation is $C_2H_5OH + 2[O] \rightarrow CH_3CO_2H + H_2O$.

Student A

(e) 2500–3300 cm^{-1}. It is an OH bond.

1680–1750 cm^{-1}. It is a C=O bond.

Student B

(e) 3000 cm^{-1}. It is an OH bond.

1700 cm^{-1}. It is a C=O bond.

ⓔ Both students gain all 4 marks, although Student A again displays better examination technique. The absorptions are listed on the datasheet supplied in the exam. Student A has used the sheet and copied out the values, while Student B has relied on memory.

ⓔ **Once again, Student A scores more than Student B (14 out of 17 compared with 9 out of 17). Student A scores 82% (a grade A), while Student B scores 53% (grade D).**

Question 22 Isomerism, alkenes, alcohols and two-stage synthesis

Time allocation: 9–11 minutes

(a) Use C_4H_8 to illustrate isomerism. Include definitions of structural isomerism and stereoisomerism. Draw structures to illustrate each type of isomerism. (4 marks)

(b) Alkenes such as propene can react with steam in the presence of an acid catalyst to produce alcohols. Explain whether or not the product would be a mixture of alcohols. (1 mark)

(c) The reaction scheme below shows some reactions of pentan-3-ol.

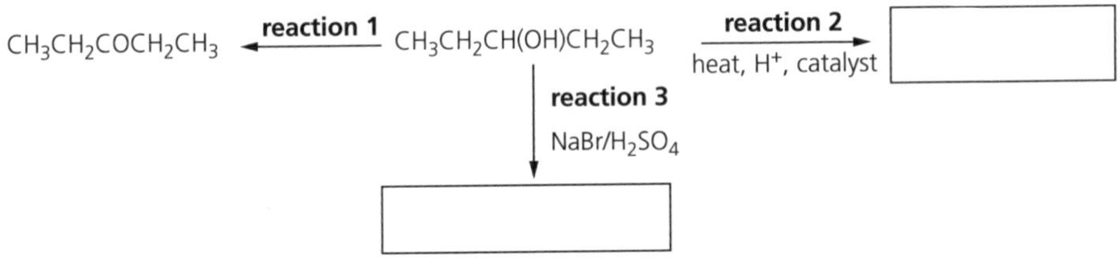

$CH_3CH_2COCH_2CH_3$ ← **reaction 1** — $CH_3CH_2CH(OH)CH_2CH_3$ — **reaction 2** (heat, H^+, catalyst) → [ ]

$CH_3CH_2CH(OH)CH_2CH_3$ — **reaction 3** ($NaBr/H_2SO_4$) ↓ [ ]

(i) State the reagents and conditions for reaction 1. (1 mark)

(ii) Draw the product of reaction 2. (1 mark)

(iii) Name the product of reaction 3. (1 mark)

(d) Propanoic acid can be prepared via a two-stage synthesis from propene. Write equations for each stage. State reagents and conditions for each step. Explain why the percentage yield is likely to be very low. (5 marks)

Total: 13 marks

e The command word 'describe' requires significant detail. When referring to a mechanism, equations are required and the movement of electrons must be tracked. In the radical substitution mechanism this is done by using a dot to track the formation and reaction of the radicals.

Student A

(a) Structural isomers have the same molecular formula but different structures. Isomers of C_4H_8 are shown below:

Stereoisomers have the same molecular formula, the same structure but different arrangement in space. But-2-ene has *cis–trans* stereoisomers.

Student B

(a) Structural isomers have the same molecular formula but different structures such as but-1-ene and but-2-ene.

Stereoisomers must have a C=C double bond, which restricts rotation, and each C in the C=C double bond must be bonded to two different atoms or groups.

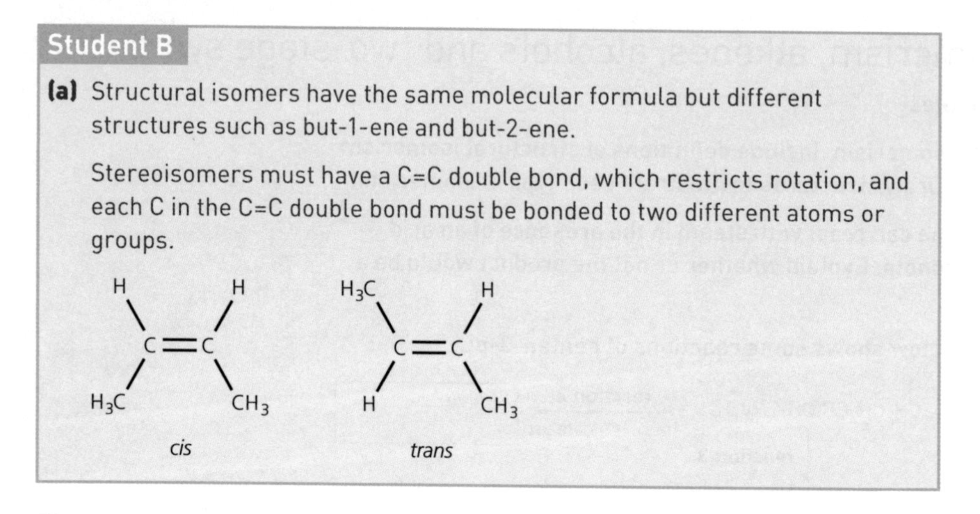

cis *trans*

ⓔ Student A gains all 4 marks, while Student B only picks up 2 marks. The definition of structural isomers is correct but the question instructs that the isomers have to be drawn *not* named. Student B's response for stereoisomers correctly explains the conditions necessary for *E*/*Z* isomerism but doesn't define stereoisomerism. Student B clearly understands isomerism but has lost two marks by not paying careful attention to the instructions within the question.

Student A

(b) $CH_3CH{=}CH_2 + H_2O \rightarrow CH_3CH_2CH_2OH$ and $CH_3CH(OH)CH_3$

unsymmetrical alkene

Student B

(b) Propene is unsymmetrical so there would be a mixture of two alcohols. Using Markownokoff's rule there would be more propan-2-ol than propan-1-ol.

ⓔ Both score the mark but Student B has wasted time by giving additional information which is correct but scores no extra marks.

Student A

(c) **(i)** acidified dichromate

(ii) $CH_3CH_2CH{=}CHCH_3$

(iii) $CH_3CH_2CHBrCH_2CH_3$

Student B

(c) **(i)** acidified dichromate and reflux

(ii) $CH_3CH_2CH(SO_4)CHCH_3$

(iii) $CH_3CH_2CHBrCH_2CH_3$

e Both score 2 marks but for different reasons. Student A has lost a mark for not stating any conditions (heat would have been sufficient). Student B has lost a mark for part (ii).

Student A

(d) $CH_3CH{=}CH_2 \xrightarrow[\text{steam, 300°C, high pressure}]{\text{step 1}} CH_3CH_2CH_2OH \xrightarrow[H^+/K_2Cr_2O_7\text{, heat}]{\text{step 2}} CH_3CH_2COOH$

You would get a poor yield because propan-2-ol would be the main product of step 1.

Student B

(d) $CH_3CH{=}CH_2 \xrightarrow[\text{steam, high temperature and pressure}]{\text{step 1}} CH_3CH_2CH_2OH \xrightarrow[Cr_2O_7^{2-}\text{, reflux}]{\text{step 2}} CH_3CH_2COOH$

The yield would be very low because each step is not 100% efficient.

e Student A scores 4 out of 5. The only mark lost is for stating 'heat' in step 2 when in order to get the carboxylic acid it must be 'heat under reflux'. Student A's explanation of the poor yield is excellent. Student B also gets 3 out of 4 for the two-stage synthesis. One mark was lost in the second step because the dichromate was not acidified. Student B's explanation of the poor yield is not specific and could apply to all reactions.

e **Yet again, Student A scores more than Student B (11 out of 13 compared with 8 out of 13). Student A scores 85% (a grade A), while Student B scores 62% (grade C).**

Extended-response questions

Questions in this section are relevant to AS component 2 and to A-level component 3.

Question 23 Trends in atomic radii

Time allocation: 6–7 minutes

The atomic radii of elements in periods 2 and 3 are shown in the table below.

		Group						
		1	2	3	4	5	6	7
Period 2	**Element**	Li	Be	B	C	N	O	F
	Atomic radius/nm	0.134	0.125	0.090	0.077	0.075	0.073	0.071
Period 3	**Element**	Na	Mg	Al	Si	P	S	Cl
	Atomic radius/nm	0.154	0.145	0.130	0.118	0.110	1.102	0.099

(a) (i) State the trend shown in atomic radius across a period. (1 mark)

(ii) Explain this trend. (3 marks)

(b) (i) State the trend shown in atomic radius down a group. (1 mark)

(ii) Explain this trend. (3 marks)

Total: 8 marks

Student A

(a) (i) Atomic radii decrease across the period.

Student B

(a) (i) It gets smaller.

e Both students gain the mark, since for these questions you do not need to write in whole sentences: 'decrease' would be sufficient.

Student A

(a) (ii) The number of protons in the nucleus increases, but the electrons fill up the same shell with the same shielding, and therefore the effective nuclear attraction increases, pulling in the outer electrons.

Student B

(a) (ii) The pull of the nucleus increases and the electrons fill up the same main shell.

e Student A gives a very good answer and scores all 3 marks. Student B clearly understands this topic, but has not used the information in the question. Three separate points are required, with each point gaining 1 mark. They are as follows: increased effective nuclear charge; electrons fill up same shell; all elements across the period experience the same degree of shielding from the inner shells.

Student A

(b) (i) Increase

Student B

(b) (i) Gets bigger

e Both answers score the mark.

Student A

(b) (ii) The number of protons in the nucleus increases, but this is compensated by the extra shell and the increased shielding by inner shells. This results in a decrease in the effective nuclear charge.

Student B

(b) (ii) There is an extra shell and more shielding, so the outer electrons are not held as tightly.

ⓔ Student A again provides a good answer and is awarded all 3 marks. Student B scores 2 marks. The essential part missing from Student B's answer is the idea that, despite the increased number of protons in the nucleus, the radius still increases due to the additional shielding.

ⓔ **Both students clearly understand the concept of variation in atomic radii, but Student B has not stated or explained fully the reasons for the variation. Student A has worked hard and has fully learnt and understood the textbook explanations, and so scores a maximum 8 out of 8 marks. Student B scores a very creditable 6 out of 8 marks, but throws away 2 marks by not taking time to reflect on what has been written in the question. Learning how to assess questions will only come with practice — it is a good idea to try timed questions as part of your revision. Don't forget that every mark counts. If Student B scored 6 out of every 8 marks throughout the exam paper, then this would translate to 45 marks out of 60, which is equivalent to a grade B.**

However, scoring 7 out of 8 marks would correspond to 52 marks out of 60, which is a grade A.

Question 24 Trends in conductivity

Time allocation: 10–11 minutes

Compare and explain the electrical conductivity of magnesium oxide, diamond and graphite. In your answer, you should consider the structure and bonding of each of the materials. (12 marks)

ⓔ This question is a 'free response' question and students have to explain the electrical conductivity of three different substances. It is a good idea to have a plan before you start. Try and work out where the marks will be allocated. By stating the conductivity, the structure and the bonding of each of the three chemicals should get you 9 marks. Explaining the conductivity by linking it to the bonding and structure should get the remaining marks.

Student A

Electrical conductivity depends on the availability of mobile charge carriers such as free electrons or free ions.

Magnesium oxide is ionic, but the ions are not free to move when solid. However, they are free to move when either molten or aqueous.

Diamond consists of a giant lattice with each carbon atom forming strong covalent bonds to four other carbon atoms. It has no free electrons and is therefore a poor conductor. Graphite also forms a giant lattice, but each carbon atom is only bonded to three other C atoms; hence it has free electrons. Graphite is a good conductor.

Student B

	Magnesium oxide	Diamond	Graphite
Conductivity	Poor when solid, good when molten or aqueous Ions only free to move when molten or aqueous	Poor No free electrons	Good Free electrons
Structure	Giant lattice	Giant lattice	Giant lattice
Bonding	Ionic	Covalent	Covalent

e Students find open-ended questions like this difficult to answer. It is advisable to plan your answer before you start. Neither student did this. The two students chose to answer in very different ways. Student A decided to write continuous prose, while Student B tabulated his/her response. Both ways of answering the question are acceptable. Student A's response is excellent, particularly the opening sentence. However, when writing in continuous prose it is easy to miss out simple statements. Student A has not stated that magnesium oxide is a poor conductor when solid but a good conductor when molten or in aqueous solution. However, Student A does imply this and so 1 of the 2 available marks is awarded. It is worth remembering, however, that examiners can only mark what is on the paper, so even the most obvious statements must be written down. Student B's approach is much more systematic and likely to lead to fewer omissions.

e **Student A scores 10 of the 11 marks. Student B gained all 11 marks.**

The mark scheme for this question is given below. Use it to mark both students' answers and see if you agree with the mark awarded. In questions like this, there are often more marking points than there are marks. When explaining the electrical conductivity of magnesium, diamond and graphite, there is a total of 13 marking points, but a maximum of 11 marks. This means that it is possible to omit 2 marking points and still gain maximum marks.

Magnesium oxide:

- giant; ionic; lattice (✓✓)
- fixed ions in solid; does not conduct when solid (✓✓)
- does conduct when in aqueous solution or molten (✓)
- mobile ions in solution or when molten (✓)

Diamond or graphite:

- covalent; giant (✓✓)

Diamond:

- no free electrons/charge carriers/all electrons involved in bonding (✓)
- does not conduct at all (✓)

Graphite:

- layered structure (✓)
- delocalised electrons (between layers) (✓)
- conducts (by movement of delocalised electrons) (✓)

13 marking points, maximum 12 marks

Question 25 Electrophilic addition

Time allocation: 6–7 minutes

Describe and explain how propene reacts with bromine. You should include in your answer:

- **any observations**
- **the name of any organic products**
- **a full description of the mechanism** (8 marks)

ⓔ The command word 'describe' requires significant detail. When referring to a mechanism, equations are required and the movement of electrons must be tracked. In the electrophilic addition mechanism this is done by showing curly arrows, relevant dipoles and lone pairs of electrons. In a free response question like this it is also useful to plan out how you are going to answer before you start.

Student A

When bromine reacts with propene it undergoes electrophilic addition. ✓ As the Br–Br molecule approaches the C=C double bond, it forms a temporary induced dipole ✓ in the Br_2, and the $Br^{\delta+}$ end is attracted to the alkene. The Br–Br bond is broken by heterolytic fission ✓ and results in a carbocation ✓. The Br^- then attacks the carbocation to form the product ✓. The product is called 1,2-dibromopropane ✓.

Student B

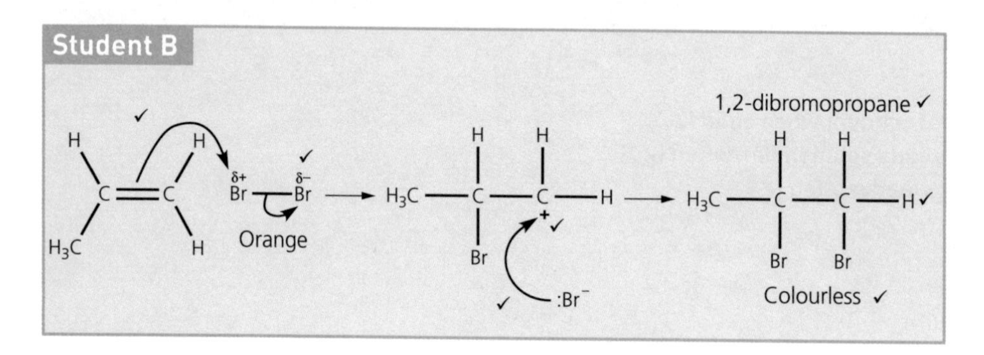

e Use the following mark scheme to mark each of the student's answers. Check your marking against the marks awarded.

- observation: (bromine is) decolorised (do not accept 'clear') (1)
- product: 1,2-dibromopropane (1)
- electrophilic addition (1)
- induced dipole in the Br_2/dipoles shown correctly on the Br–Br bond and curly arrow on Br–Br bond as shown/heterolytic fission (1)
- curly arrow from the π-bond to the bromine, or words to that effect (1)
- intermediate carbonium ion/carbocation (1)
- curly arrow from Br^- back to the carbonium ion/carbocation/nucleophilic attack/Br^- forms a covalent bond with the carbocation (1)
- lone pair of electrons shown on the Br^- (and curly arrow from lone pair to the carbonium ion/carbocation)/Br^- acts as a lone pair donor (1)

Students often find free-response questions difficult to answer. In the question you are asked to state any observations and name any organic products and you are informed that 1 mark is available for the quality of written communication. This accounts for 3 of the available 9 marks, leaving 6 marks for the mechanism; therefore six separate points are required.

Both students have tackled this question well, but differently. Student A has written in prose and described the mechanism well. Describing a mechanism this way is difficult and requires a great deal of thought, planning and care. Student A clearly understands the mechanism, but has failed to show or to describe in words, the movement of electrons. He/she has concentrated so hard on describing the mechanism that the observation has been omitted. This is easily done.

Student B has taken the opposite approach — few or no words — and many students communicate like this. Student B has scored 5 of the 6 marks for the mechanism as well as the marks for stating the observation and naming the organic product. However, he/she has failed to state that the mechanism is an electrophilic addition mechanism, and hence loses 1 mark. Examiners accept that chemists communicate in a variety of ways: equations, diagrams and tables. None of these is penalised.

e Both students score 7 out of 9 marks. Strangely, Student A has attempted to write all of the answer in continuous prose. This is extremely difficult, and a simple observation mark has been lost. Student B has simply ignored the need for continuous prose, but has scored well. It is important to try to strike the right balance.

Question 26 Analytical techniques

(a) Compound Z contains 62.1% C, 10.3% H and 27.6% O by mass. Use this and the spectra below to identify compound Z. Show all of your working. (8 marks)

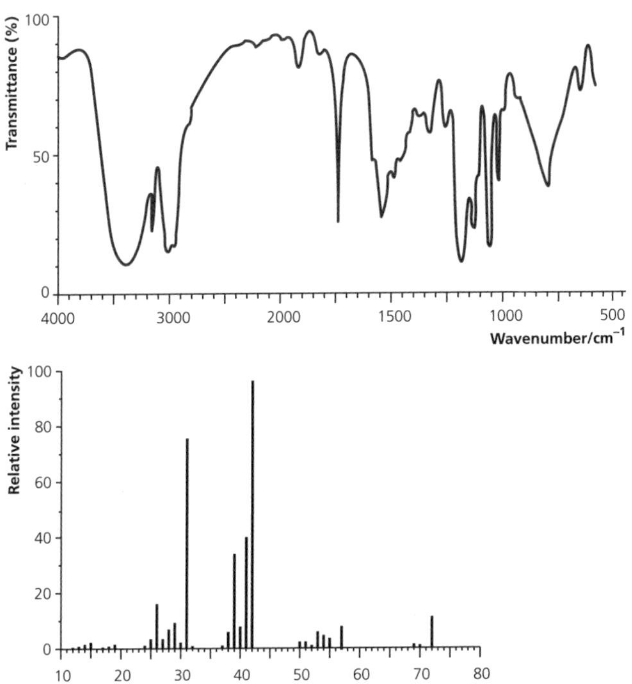

(b) Draw the product that would be formed when compound Z reacts with:

(i) acidified potassium dichromate under reflux

(ii) acidified potassium dichromate under distillation

(iii) steam in the presence of an acid catalyst (3 marks)

If you are unable to identify compound Z in part (a) use but-2-en-1-ol ($CH_3CH=CHCH_2OH$) in place of compound A.

Total: 11 marks

e This question is open-ended and there is no set way of answering part (a). It is essential that you use all the information in the question and in order to score 8 marks you must make 8 separate points. Part (b) is much more structured and you must follow the command word and 'draw' the organic products.

Student A

(a) Empirical formula C:H:O

(62.1/12):10.3:(27.6/16)

5.175:10.3:1.725

3:6:1

Empirical formula is C_3H_6O, hence empirical mass = 58

From the mass spectrum the molecular ion peak is at m/z = 58, hence the molecular formula is also C_3H_6O

The infrared spectrum has a broad peak in wavenumber range 3200–3600 cm^{-1} which shows an alcohol is present hence C_3H_5OH.

The alkyl group, C_3H_5 is unsaturated which is confirmed by the peak in the infrared spectrum for the C=C bond in the range 1620–1680 cm^{-1}.

The peak at m/z = 31 in the mass spec indicates a primary alcohol fragment ($CH_2OH^+(g)$)

The possible isomers are:

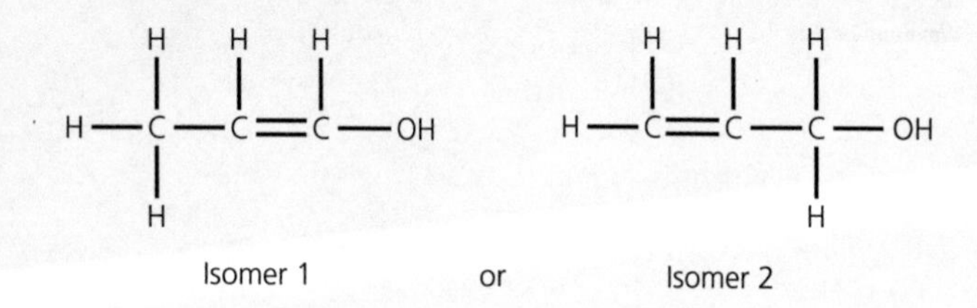

but only isomer 2 would give a peak at 31 so it must be isomer 2.

Student B

(a) Empirical formula C:H:O

(62.1/12):10.3:(27.6/16)

5.175:10.3:1.725

3:6:1

Empirical formula is C_3H_6O, hence empirical mass = 58

From the mass spectrum the molecular ion peak is at m/z = 58, hence the molecular formula is also C_3H_6O

The infrared spectrum has a broad peak in wavenumber range 3200–3600 cm^{-1} which shows an alcohol is present hence C_3H_5OH.

It must be:

OH
CH
H_2C—CH_2

e Use the following mark scheme to mark each of the student's answers. Check your marking against the marks awarded.

- empirical formula (1)
- use mass spec to find molar mass (1)
- molecular formula (1)
- uses infrared to identify OH group (1)
- uses infrared to identify C=C group (1)
- uses mass spec to identify primary alcohol $CH_2OH^+(g)$ (1)
- considers all possible isomers (1)
- identifies compound Z from their workings (1)

Both students have tackled this question well. Student A's answer is brilliant and has scored full marks. Student A has been really logical and has used all the information given in the question.

Student B has also done well and scores the first 4 marking points. Unfortunately Student B has failed to spot the peak for the C=C bond in the infrared spectrum and also hasn't made use of any of the fragment ions in the mass spectrum. This has cost Student B the next two marking points. However, he/she would get a mark ✓ ecf (error carried forward) for identifying a compound that matches their workings. Student B might also get the mark for considering all isomers as there is only one possible saturated alcohol with formula C_3H_6O.

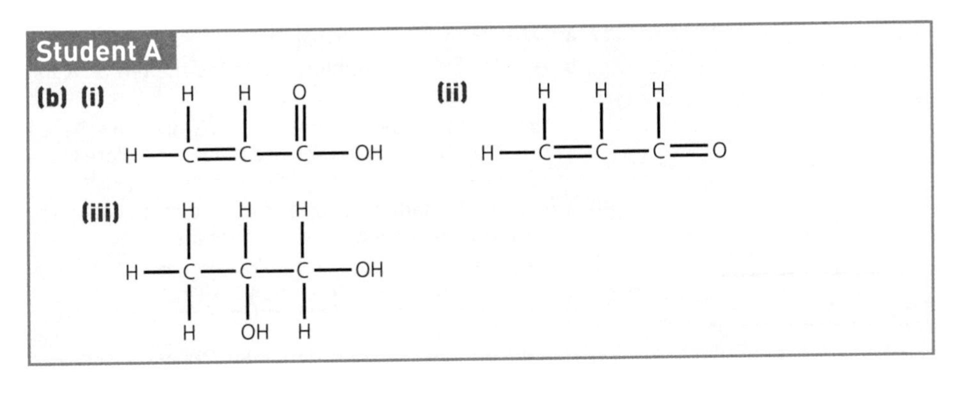

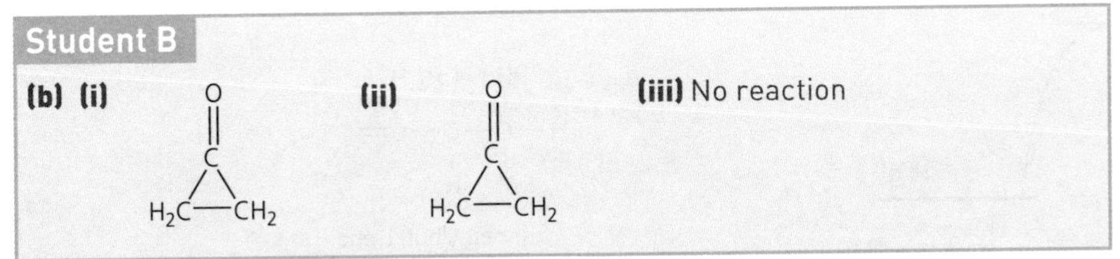

e Student A has done very well again and scored all 3 marks. Student B also gets full marks but all 3 marks are awarded as ✓ ecf from the organic product identified in part A.

e **Student A has scored maximum marks. Student B has scored a very creditable 8 out of 11 (possibly 9) and shows that persistence and carefully following the instructions can lead to marks even when errors are made.**

Knowledge check answers

1 The correct answer is B as across a period melting point increases from group 1 to group 4. This is followed by a large drop in groups 5, 6 and 7.

2 atomic radii Al < Mg < Na < K
conductivity K < Na < Mg < Al
ionisation energies K < Na < Mg < Al

3 $Sr + 2H_2O \rightarrow Sr(OH)_2 + H_2$
Reactivity increases down the group. ✓
When (group 2) metals react they lose electrons ✓ and electrons are easier to remove as we go down the group because atomic radius increases ✓ and shielding increases ✓ which together outweigh the increase in the nuclear charge. ✓

4 Reactivity decreases down the group.
When (group 7) non-metals react they gain electrons. ✓ The increase in atomic radius ✓ and the increased shielding ✓ reduce the attraction for electrons, which together outweigh the increase in the nuclear charge. ✓

5 Activation energy is the minimum energy required to start a reaction.

6

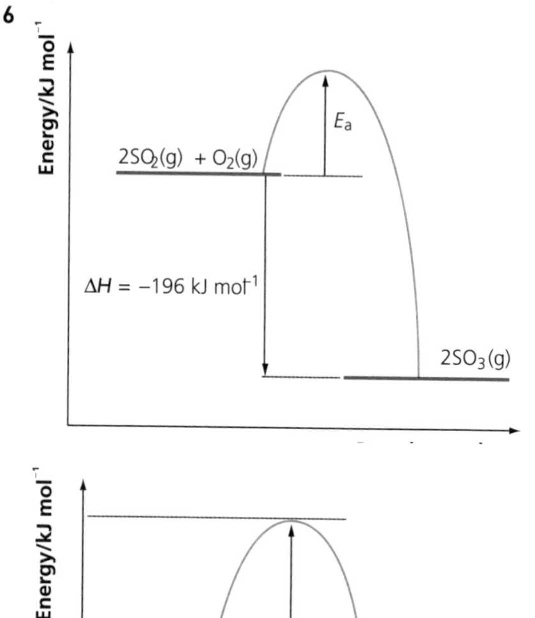

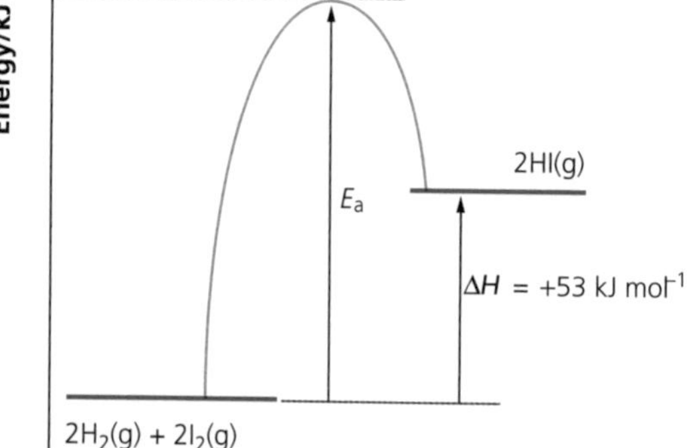

7 The enthalpy change when 1 mole of substance is formed from its elements under standard conditions of 298 K and 101 kPa.
$3C(s) + 3H_2(g) + \frac{1}{2}O_2(g) \rightarrow CH_3COCH_3(l)$

8 The enthalpy change when 1 mole of substance is burnt in excess oxygen under standard conditions of 298 K and 101 kPa.
$CH_3CH_2CHO(l) + 4O_2(g) \rightarrow 3CO_2(g) + 3H_2O$

9 $q = mc\Delta T = 500 \times 4.18 \times 2205 = 47025\text{ J} = 47.025\text{ kJ}$
$\Delta H = q/n = 47.025/(2.48/62) = 1175.6\text{ kJ mol}^{-1}$

10 -125 kJ mol^{-1}

11 -79.2 kJ mol^{-1}

12 Increasing pressure forces the gaseous particles closer together (volume decreases) and increases the chance of a collision.

13 The new curve goes through the origin, the peak is higher than the original and to the left (lower energy) of the original, at high energy the new curve is below the original curve.

14 Le Chatelier's principle states that if a closed system under equilibrium is subject to a change, then the system will move in such a way as to *minimise* the effect of the change.

15 **a** moves to the **i** left, **ii** right, **iii** right, **iv** no change; **b** heterogeneous

16 $K_c = [N_2O_4(g)]/[NO_2(g)]^2$
a K_c is small so the equilibrium lies to the left.
b $0.0025 = [N_2O_4(g)]/[1]^2$ therefore $[N_2O_4(g)] = 0.0025\text{ mol dm}^{-3}$

17 **a** $K_c = [H_2(g)][CO(g)]/[H_2O(g)]$
b **i** At 25°C the equilibrium lies to the left as K_c is very small.
ii As the temperature increases K_c increases, hence the equilibrium moves to the right, therefore the forward reaction must be endothermic, $+\Delta H$.

18 **a** 2-bromopropane **b** but-2-ene **c** 2,2-dimethylpropane or dimethylpropane **d** 3-methylpentane

19

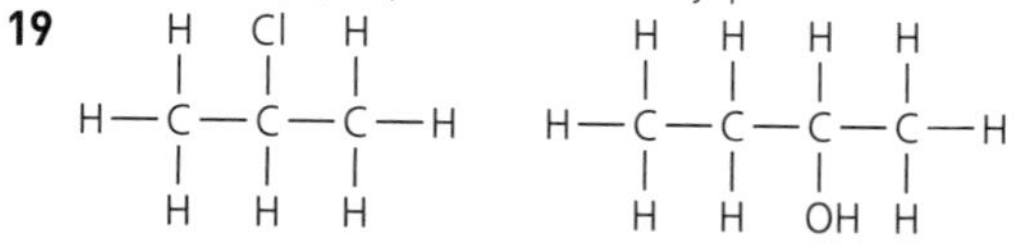

2-chloropropane

butan-2-ol

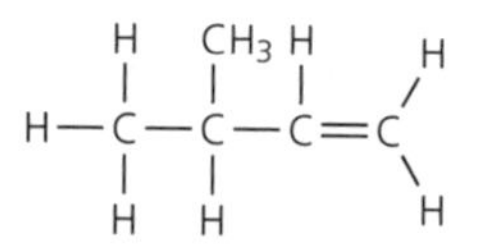

3-methylbut-1-ene

20

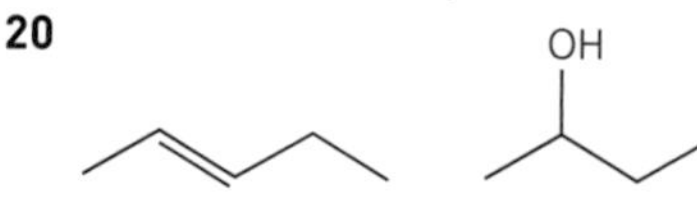

pent-2-ene butan-2-ol

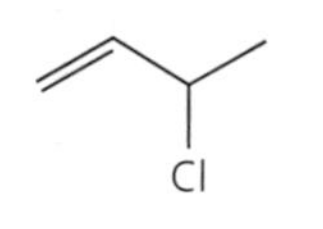

3-chlorobut-1-ene

21 moles of methanol = 5.0/32 = 0.156, moles of ethanoic acid = 9.38/60 = 0.156

moles of methyl ethanoate = 5.2/74 = 0.0703

a percentage yield = (0.0703/0.156) × 100 = 45%

b atom economy = [74/(74 + 18)] × 100 = 80.4%

22 2,3-dimethylbutane, 2-methylpentane, hexane, octane

23 There are five possible isomers '*x*'-bromo-2-methylpentane where '*x*' can be 1, 2, 3, 4 or 5.

24

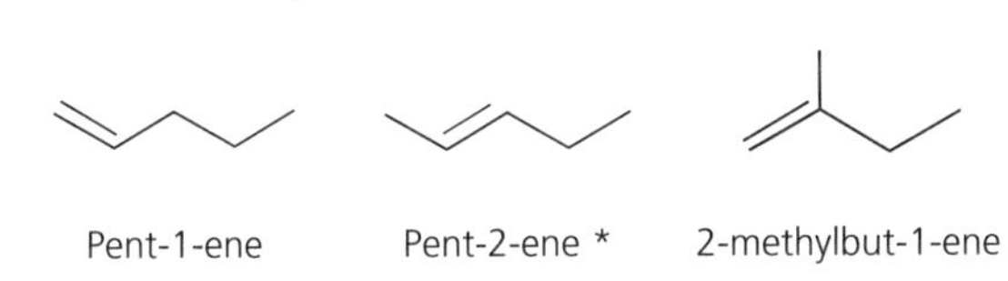

Pent-1-ene | Pent-2-ene * | 2-methylbut-1-ene | 2-methylbut-2-ene | 3-methylbut-1-ene

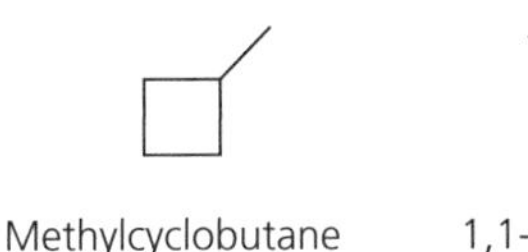

Cyclopentane | Methylcyclobutane | 1,1-dimethylcyclopropane | 1,2-dimethylcyclopropane | Ethylcyclopropane

Pent-2-ene * also has *E*/*Z* isomers

25 $H_2C{=}CHCH{=}CH_2 + 2Br_2 \rightarrow CH_2BrCHBrCHBrCH_2Br$; the Br_2 would be decolorised. Organic product: 1,2,3,4-tetrabromobutane.

26 a An electrophile is an electron-pair acceptor.

$C_6H_{10} + Br_2 \rightarrow C_6H_{10}Br_2$

or

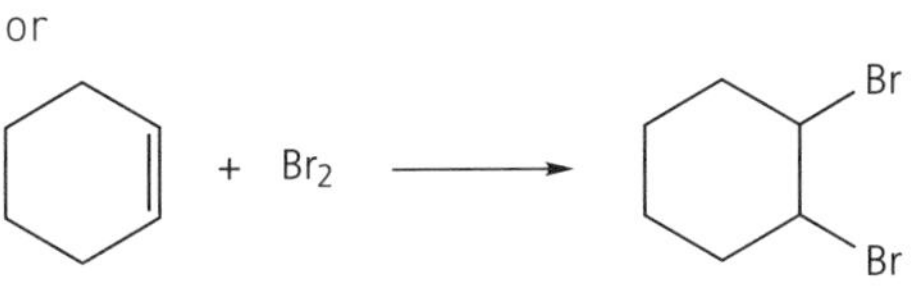

b

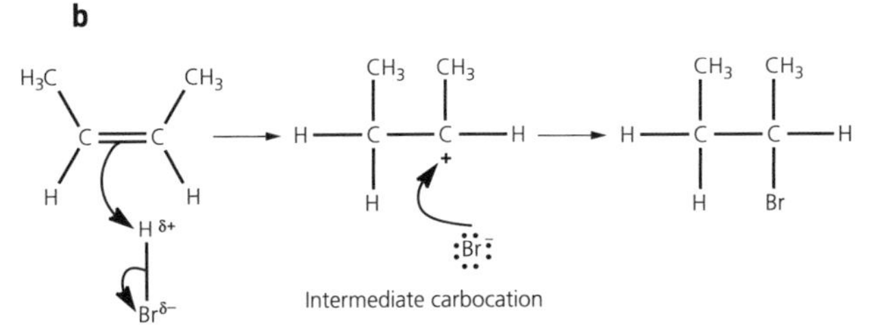

27 2-methylpropan-2-ol would be the major product as it would be formed via the more stable intermediate which is a tertiary carbonium ion.

28

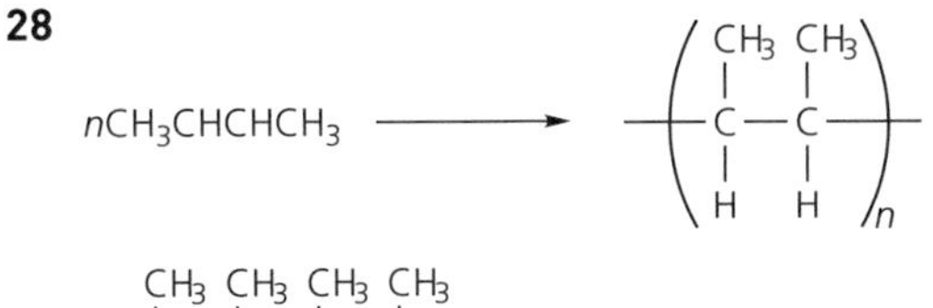

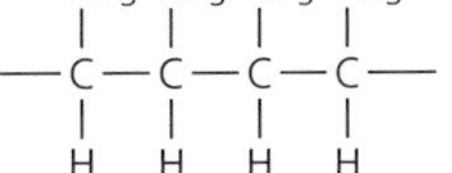

29 a

i $C_3H_7OH + 4\tfrac{1}{2}O_2 \rightarrow 3CO_2 + 4H_2O$

ii $CH_3CH_2CH_2OH \rightarrow CH_3CH{=}CH_2 + H_2O$

iii $CH_3CH_2CH_2OH + 2[O] \rightarrow CH_3CH_2COOH + H_2O$

iv $CH_3CH_2CH_2OH + [O] \rightarrow CH_3CH_2CHO + H_2O$

v $CH_3CH_2CH_2OH \xrightarrow[H_2SO_4]{NaBr} CH_3CH_2CH_2Br$

b Reflux: continuous evaporation and condensation such that volatile components cannot escape. Distillation: evaporation followed by condensation such that volatile components can escape.

30 a $CH_3CH_3 + Cl_2 \xrightarrow{UV} CH_3CH_2Cl + HCl$

$CH_3CH_2Cl + {}^{-}OH \xrightarrow{heat} CH_3CH_2OH + Cl^-$

b $CH_3CH_2OH \xrightarrow[170°C]{H_2SO_4} CH_2{=}CH_2 + H_2O$

$CH_2{=}CH_2 + H_2 \xrightarrow[150°C]{Ni\ cat.} CH_3CH_3$

31 CH_3^+, $CH_3CH(OH)^+$, $CH_3CH(OH)CH_2^+$, $CH_3CH(OH)CH_2CH_3^+$

32 molar mass of $CH_2{=}CHCH_2CH_2OH$ is $72\,g\,mol^{-1}$

Mass spectrum: $m/z = 72 = CH_2{=}CHCH_2CH_2OH^+(g)$; $m/z = 31 = CH_2OH^+(g)$

Infrared spectrum: $3200–3600\,cm^{-1}$ O–H; $1620–1680\,cm^{-1}$ C=C.

Note: **bold** page numbers indicate defined terms.